Minnie Vaid has juggled multiple roles over a three-decade stint in mainstream media. She is a print and television journalist, a documentary filmmaker, creative producer for feature films and more recently, author of three non-fiction books, *A Doctor to Defend: The Binayak Sen Story* (2011); *Iron Irom: Two Journeys* (2013) and *The Ant in the Ear of the Elephant* (2016). Her areas of interest include social and political issues in rural India, human rights, the environment and gender.

ALSO BY MINNIE VAID

A Doctor to Defend: The Binayak Sen Story (2011)
Iron Irom: Two Journeys (2013)
The Ant in the Ear of the Elephant (2016)

THOSE
Magnificent
WOMEN
and Their
FLYING
MACHINES

ISRO'S MISSION TO MARS

MINNIE VAID

SPEAKING
TIGER

SPEAKING TIGER PUBLISHING PVT. LTD
4381/4, Ansari Road, Daryaganj
New Delhi 110002

Copyright © Minnie Vaid 2019

ISBN: 978-93-88874-58-8
eISBN: 978-93-88326-89-6

10 9 8 7 6 5 4 3 2 1

The moral rights of the author have been asserted.

Typeset in Sabon Roman by SÚRYA, New Delhi

All rights reserved.
No part of this publication may be reproduced, transmitted,
or stored in a retrieval system, in any form
or by any means, electronic, mechanical,
photocopying, recording or otherwise,
without the prior permission
of the publisher.

This book is sold subject to the condition that it shall not, by way
of trade or otherwise, be lent, resold, hired out,
or otherwise circulated, without the publisher's
prior consent, in any form of binding
or cover other than that in
which it is published.

Contents

List of Abbreviations	ix
Mars Orbiter Mission: Timeline	xi
Prologue	1
1. Women from Mars	7
2. MOM: Operations in Outer Space	25
3. MOM: The Payload Performers	84
4. The Vanguard Veterans	105
5. Beyond MOM: The Applications Achievers	140
6. Crossing the Rubicon	175
Epilogue	206
Acknowledgements	209
Notes	210
References	217

For my parents,
Dr Jawaharlal Vaid and Mrs Prem Vaid

List of Abbreviations

ASLV: Augmented Satellite Launch Vehicle
DST: Department of Science and Technology
GRB: Gamma Ray Burst
IIA: Indian Institute of Astrophysics
INSAT: Indian National Satellite System
IISc: Indian Institute of Science
IISU: ISRO Inertial Systems Unit
IPRC: ISRO Propulsion Complex
ISTRAC: ISRO Telemetry, Tracking and Command Network
JAXA: Japan Aerospace Exploration Agency
LPSC: Liquid Propulsion Systems Centre
MCF: Master Control Facility
NRSC: National Remote Sensing Centre
PSLV: Polar Satellite Launch Vehicle
SDSC: Satish Dhawan Space Centre
SAC: Space Applications Centre
SHAR: Sriharikota Range
SSTC: Space Science and Technology Centre
SROSS: Stretched Rohini Satellite Series
TERLS: Thumba Equatorial Rocket Launching Station
URSC: U.R. Rao Satellite Centre
VSSC: Vikram Sarabhai Space Centre

Mars Orbiter Mission

Timeline

5 November 2013: PSLV-C25 successfully launches Mars Orbiter Mission Spacecraft from SDSC SHAR, Sriharikota.

7 November 2013: The first orbit-raising midnight manoeuvre of the Mars Orbiter Spacecraft successfully completed. The spacecraft has to go several rounds around earth to gradually increase its velocity to attain the escape velocity with minimum fuel consumption. This is done in a series of midnight manoeuvres (Earth Bound Manoeuvers) in which MOM's engine is fired in a direction tangential to earth while MOM is at its closest orbital position to earth.

8 November 2013: The second orbit-raising manoeuvre of the Mars Orbiter Spacecraft successfully completed.

9 November 2013: The third orbit-raising manoeuvre of Mars Orbiter Spacecraft successfully completed.

11 November 2013: In the fourth orbit-raising operation, the apogee (farthest point from earth), of the Mars Orbiter Spacecraft was raised from 71,623 km to 78,276 km. The spacecraft was in normal health.

12 November 2013: The fourth supplementary orbit-raising manoeuvre of the Mars Orbiter Spacecraft successfully completed.

16 November 2013: The fifth orbit-raising manoeuvre of Mars Orbiter Spacecraft successfully completed.

1 December 2013: Trans-Mars Injection (TMI) operations completed successfully. After this manoeuvre, the earth-orbiting phase of the spacecraft ended. It is now on course to Mars after a journey of about ten months around the sun.

2 December 2013: The spacecraft has travelled a distance of 5,36,000 km.

4 December 2013: The spacecraft has traversed beyond the Sphere of Influence (SOI) of the earth, going beyond 9,25,000 km.

11 December 2013: The first Trajectory Correction Manoeuvre (TCM) of the spacecraft is carried out successfully. MOM is at a distance of about 29 lakh (2.9 million) km away from earth.

9 April 2014: The Mars Orbiter Spacecraft crosses the half-way mark of its journey.

16 September 2014: Time-tagged commands to execute the Mars Orbit Insertion (MOI) uploaded.

17 September 2014: Uploading of commands for the Fourth Trajectory Correction Manoeuvre and test firing of the Main Liquid Engine (scheduled for 22 September) in progress.

22 September 2014: Test firing of Main Liquid Engine of Mars Orbiter Spacecraft is successful.

24 September 2014: Mars Orbiter Spacecraft enters Mars orbit.

Prologue

Fade in: Santo, a young girl of twelve, is reading a book about Mars in her home in Rewari district, Haryana. A visitor asks her what she wants to be when she grows up—a doctor or an engineer. 'Astronaut,' she replies softly. The man says, 'In our community those who are able to get to Delhi are considered successful. If they reach London or America, they become examples to be followed. First get to Delhi, then think about sitting in a rocket and reaching the moon.'

The little girl's face falls.

Over the next few days, even as her mother admonishes her—'Your final exams are approaching and you're fixated about Mars?'—she paints her father's helmet astronaut-white, calling it her Mars helmet, makes a model rocket and installs homemade star lights in her room. Her mother tells the father they should talk Santo out of her childish phase or she won't do well in her exams. The father believes that this could be her dream and she should be encouraged. They decide to gift her a laptop. The mother hugs her saying, 'You have to become Rewari's dream, use the laptop as a rocket and fly off to Mars. My astronaut.'

The tagline, GIFT THEM CURIOSITY, GIFT THEM DREAMS, appears on the screen and the video ends. This Lenovo ad, championing the girl child, skilfully subverts age-old stereotypes. In doing this, it provides hope for the future of the girl child, all in less than three minutes.[1]

If only real life were that simple.

Those Magnificent Women and Their Flying Machines

One hundred and thirty-two years ago, when Dr Anandibai Joshi became India's first female physician with an MD degree from an American medical college, she could not have imagined that the women following her example would be facing similar battles more than a century later. A few years after Dr Joshi, in 1933, Dr Kamala Sohonie was denied admission to the Indian Institute of Science (IISc) in Bengaluru by Nobel Laureate C.V. Raman, solely because of her gender. It was only her persistence which enabled her to complete three years at IISc and move on to Cambridge University to receive a doctorate in science.

How many of us have heard of Janaki Ammal, Anna Mani, Asima Chatterjee, Rajeshwari Chatterji, Charusita Chakravarty and Mangala Narlikar? These pioneering women scientists were a mix of physicians, botanists, chemists and physicists working against formidable odds to carve their place in the history books. Their contributions are mostly known within the academic community or to students researching gender in Indian science in the early 1990s.

The sex ratio at the IISc, a premier institute for scientific research in India, has risen from 2 per cent in the 1960s, i.e. two female students for 100 male students, to 19 per cent in 2016. The progress hasn't quite been meteoric.

Every woman learning, teaching or practicing science in India has her own unique set of challenges, even with predecessors and role models providing hope on a difficult path.

Here are a few indicators of how tough it is to be a female scientist, not just in India:

- Women make up only 28.8 per cent of those employed in scientific research and development across the world.[2]
- They are less likely to enter and more likely to leave careers in STEM (Science, Technology, Engineering and Mathematics).[3]
- They are poorly represented in science academies—there are only 12 per cent female members in sixty-nine science academies worldwide.[4]
- Only seventeen women have been awarded a Nobel Prize in the three science categories since the award's inception in 1901.[5]
- The latest joint winner of the Nobel Prize for physics in 2018, Canadian Donna Strickland, is the first female laureate in fifty-five years and only the third woman to win in physics. Her short Wikipedia page was created *after* she received the Nobel.[6]

Closer home, India's top science prize, the Shanti Swarup Bhatnagar award has been given to only sixteen women out of a total of more than 500, since its inception in 1958.[7]

Women also receive less than 5 per cent of the fellowships awarded by the three major national science academies,[8] are quoted less often, rarely invited as speakers at plenary science conferences and hardly ever head advisory committees or science academies.[9] Only one out of the Indian National Science Academy's forty-one past office bearers was a woman, and just fourteen out of INSA's 501 awards were given to women.[10]

Those Magnificent Women and Their Flying Machines

Most importantly, there is a major dearth of women in leadership positions, as heads of scientific centres and organizations, research institutes or in higher decision-making committees. Unconscious or implicit biases limit women's progress in scientific and engineering fields.

Numerous social and environmental factors cause the obvious disparity between the numbers of male and female scientists, not just in India but also across the world. Gender roles prime women to assume the lion's share of domestic responsibilities from an early age, along with manoeuvring the work-family balance. Coping strategies such as ensuring a support system at home with in-laws or domestic help, flexible timings, and working in the early hours are principally applicable to the woman scientist. Finding time to do science, which is not a nine-to-five job, or putting in longer hours at the lab/office to make that breakthrough comes at a high cost. Many women scientists limit themselves to less challenging positions, stopping short of jockeying for higher posts, which involve travelling, to ensure they have sufficient time and energy to perform the other roles expected of them. This has its inevitable effect on recruitment. Both male and female scientists interviewed for this book affirm that, all things being equal, a male candidate is often preferred over a female. Such biases, which position men as 'born leaders', set them on the path of career success while leaving women on the sidelines, or making choices that are seen to be easier. Those women who consciously choose and work hard at building successful careers in science are considered trailblazers for a new generation of girls, for whom gender will be irrelevant someday.

Pursuing a career in science involves at least eight to nine years of studying; first for a BSc or BTech degree, followed by post-graduation and doctorate studies. Working in the private sector, at a reputed pharmaceutical company for example, a fresh entrant with a PhD would be paid approximately Rs 67,000 per month. As you climb higher up the corporate ladder, the pay scale ranges from Rs 40–50 lakhs per annum for scientists with fifteen years of experience, while a vice-president's salary would be over Rs 1 crore, along with stock options. The topmost positions such as that of the president of a business unit division (PhD and post-doctorate is mandatory for such posts) would command over Rs 2.5 crores per annum.

A PhD is mandatory for the private sector but not for government organizations such as the Indian Space Research Organisation (ISRO), where science and engineering graduates are selected after a written test and an interview.

Today a newly recruited scientist in ISRO gets a monthly salary of Rs 67,800 (after probation, this is approximately Rs 70,000), inclusive of transport, house rent allowance, insurance, medical facilities and pension. If she makes it right to the top, the chairperson's monthly salary would be Rs 2,50,000 along with benefits.

No woman has reached the top rank at ISRO since its inception on 15 August 1969. However, the last fifty years have seen an increasing number of women scientists and engineers take on leadership roles as project managers, project directors and programme directors. They have carved their own space in a fiercely competitive, male-dominated world, managing homes and families through meticulous planning, efficient organization and an unflappable temperament. These are essential ingredients of the scientific temperament, required to excel in any kind of science—especially in space research.

They wear the gender tag lightly—they don't call attention to it, they don't confront discriminatory mindsets with any aggression. They simply let their work speak. And only once in a while, the rest of the country listens and watches in appreciation. Nowhere was this more evident than in the accolades given to the women scientists of the Mars Orbiter Mission (MOM) in September 2014.

Meet Mangalyaan's 'superwomen' along with ISRO's pioneering women scientists as well as the next generation leaders—this is their remarkable story.

Chapter 1

Women from Mars

I have frequently been asked how I could reconcile family life with a scientific career. It has not been easy... We must believe that we are gifted for something and that this thing must be attained.

—Marie Curie, the first woman to receive the Nobel Prize in 1903 and the first to receive it twice in two different sciences: physics and chemistry

24 September 2014, 6.56 a.m., twenty-one minutes before the Mars Moment of Insertion burn (i.e. firing of the engine) at Peenya (near Bengaluru):

It's almost exactly as depicted in science fiction films. Months of non-stop, frenetic activity and meticulous preparations culminating in a nail-biting finale watched by millions of viewers across the world.

For the mission team however, the magic unfurls right in front of their eyes, in a special place quite their own. In the vast Mission Operations Complex (MOX 2) at ISTRAC (the ISRO Telemetry, Tracking and Command network), there is pin-drop silence. All eyes are fixed on the drama unfolding in front of their eyes in real time, on gigantic video screens and individual computer terminals.

The huge room, with its own special VIP viewing gallery on the first floor, is divided into concentric semicircles. There are six 54-inch video screens mounted high above in a row, four rows of individual computer terminals manned by experts, flanked at the outer edges by assistants, with the core operations scientists of the Mars Mission Team seated at the centre—their hearts and minds racing. Among the 300-odd scientists and engineers in MOX2, sit two (among at least ten other) women, sharing the enormity and excitement of the day with great fervour. It is, after all, a culmination of their fascinating eighteen-month journey, from concept to execution.

~

Nandini Harinath, the petite, sari-clad 47-year-old deputy operations director and project manager, mission design, of the Mars Orbiter Mission (MOM), describes the run-up to that unforgettable moment of truth: 'We had to identify all possible scenarios and work out all mitigation plans within a small time-span of eighteen months. The Mars Moment of Insertion (MOI) is a one-time event. There were no second chances. We had to demonstrate that our satellite is capable of going into the Martian atmosphere and orbiting around.'

Similarly, Ritu Karidhal, the bespectacled 42-year-old deputy operations director and project manager, operations, MOM, shares the enormity of the tasks involved: 'For the Mars interplanetary mission there had to be 99.99 per cent accuracy in space. There was absolutely no margin for error, plus we had no earlier heritage of knowledge either. It was a totally new project. And being mission design and operations directors, we had to foresee the challenges in managing the mission itself—from when you leave the earth to when you enter the Mars orbit. We had to design a spacecraft that was smart, autonomous and could troubleshoot on its own. And we had to do it in a specific time-frame.'

~

Almost 1,500 km away in Ahmedabad, as a packed auditorium of scientists are glued to their screens, two other women, who have spent the better part of two years in windowless 'clean rooms'—not knowing or caring if it was daylight outside—working towards this path-breaking moment, are keeping their fingers crossed.

Moumita Dutta, the 39-year-old project manager for the methane sensor of the MOM, says, 'For one and a half years I had only Mars on my mind. Every day I would think of new test set-ups that I was developing: Will they work or not? Will they give me the desired performance? Will the instruments work? Then sourcing all the components, testing and integrating them, aligning them precisely, right down to the last micron...'

Minal Sampat, the affable 39-year-old project manager, systems integration, MOM, outlines her own contribution: 'Working in a soundproof, pressurized "clean room" without sunshine or any other external noise for days on end—often without a break during testing—was a challenge, but the payloads [scientific instruments onboard the spacecraft, see n. 20] are like my babies. I could not leave them.'

~

Those Magnificent Women and Their Flying Machines

As the countdown for the MOM satellite to enter the Mars orbit begins, after 300 days of circling in space, the mood on earth at the various ISRO centres across the country is a mix of apprehension and quiet confidence. The confidence is not misplaced, given ISRO's proven track record of successful spacecraft missions—from the very first experimental Aryabhata in April 1975 to the Mars Orbiter lift-off on 5 November 2013.

However, MOM stands out among all the previous missions of ISRO.

It is the first time that the Indian space agency is venturing into interplanetary travel—exploring Mars, the Red Planet, which has fascinated not just space scientists, but all of mankind. Is there life on Mars? Do Martians really exist beyond science fiction novels and space fantasy films? Is the core of Mars solid, liquid or does it have two sub-layers like that of the earth? How did Mars lose the water it once had? Some of these questions have puzzled our scientists and space enthusiasts for decades. The main reason for this preoccupation is the widely-held belief that despite no recorded signs of civilization, Mars is possibly the only planet where human beings could live one day.

Recently, a huge lake of liquid water *was* found on Mars by astronomers using the Mars Advanced Radar for Subsurface and Ionosphere Sounding (MARSIS) onboard the Mars Express Orbiter.[11]

Unlike the earth, Mars is a hostile region with toxic soil and abundant radiation (the lack of a preventive ozone layer exposes it to harmful ultra-violet rays of the sun). Scientists describe Mars—the fourth planet from the sun and the second closest to the earth—as a cold desert world, half the earth's diameter, with the same amount of dry land. It has less gravity than the earth—a person on earth would weigh 62 per cent less on Mars.[12] Mars also has seasons, polar ice caps, volcanoes and canyons. The atmosphere on Mars is a hundred times less dense and largely composed of carbon dioxide, so one needs to breathe 14,500 breaths to get the volume of oxygen equivalent to one earth-breath. Mars is also six times smaller than the earth. One day on Mars, also known as 'sols', is thirty-seven minutes longer than the 24-hour-day on earth.

History of Mars Expeditions

The gap between existing, proven knowledge about Mars and its unexplored and unexplained phenomena is wide enough to have inspired regular space journeys over the years. These range from Russian spacecrafts from 1957 and 1971, as well as NASA's American Mariner 9 in 1971, the American Viking mission in 1976, the Pathfinder Sojourner and Viking Landers in the 1990s, to the Curiosity Rover landing in 2012, Maven in 2013 and India's MOM in 2014. The Orbiters, Landers and Rovers have paved the way for a manned mission to Mars in the near future, with further multi-nation expeditions aiming for human settlement planned for 2035–40. A population of 150–180 would allow normal reproduction for 60–80 generations, equivalent to 2,000 years, according to veteran ISRO scientist S. Adimurthy.[13]

As the mystique around the planet continues to grow, the expeditions have found fun ways to involve the general public as well. The 'Send Your Name to Mars' campaign by NASA in November 2017 received a phenomenal 2,429,807 submissions from enthusiasts across the world. Americans, Chinese and Indians topped the list.[14]

All these people wanted to leave their imprint, quite literally, on the Red Planet. Their names, etched on a silicon wafer microchip through an electron beam which forms letters with lines one-thousandth the diameter of a human hair, took off on NASA's InSight robotic lander deck on 5 May 2018. This chip, attached to the Insight Lander deck, will remain on Mars forever.

A few years earlier, more than 100,000 people had signed up for a one-way trip to Mars on the ambitious Mars One programme aiming to colonize the Red Planet in 2022.[15]

In February 2018, SpaceX chief Elon Musk's red Tesla Roadster with 'Starman', a mannequin dressed in a spacesuit onboard, was launched successfully by the SpaceX Falcon Heavy rocket towards Mars. The car ended up as a fly-by, and is now headed for the asteroid world.[16]

NASA has also employed researchers to design robotic bees—sensor-fitted fast-moving micro-bots— that can fly on Mars, mapping its terrain and collecting samples of its thin air for traces of methane gas, which is considered a potential source of life.[17]

Mars in Popular Culture

What is it about Mars that has fascinated people across continents for centuries? In Vedic beliefs and mythology,

Women from Mars

Mars, also known as Mangal, Bhauma, Angaraka, Chevaai and Kuja, is supposed to be born of Bhumi, the Goddess Earth. Mangal represents war and warriors, violence and valour, ambition and fire. Numerous myths validate Mars as the protector of dharma, the sacred path of righteous life. A prayer offered to this planet guarantees freedom from poverty and illness.

In vivid contrast to this image of Mars as the protector, Orson Welles' 1938 CBS radio broadcast *War of the Worlds* scared millions of listeners into believing that Martians had invaded earth in their war-machines. Welles described them as 'strange beings who descended out of a spacecraft in a farm in New Jersey as the vanguard of an invading army from the planet Mars' with 'tentacles, black eyes that gleamed like serpents, a V-shaped mouth with saliva dripping from rimless lips...a monster weighed down by gravity'.[18] Panicked listeners flooded the roads, hid in basement cellars and loaded their guns to overcome the alien invaders, before finally realizing that the broadcast was a Halloween gag. More recently, the hugely successful 2015 film *The Martian* tapped into popular fantasies of how a man stranded on Mars would survive, and whether he could ever return home safe and sound.

India's Mars Mission

At its closest, Mars is still 54.6 million km away from earth.

As the earth is closer to the sun, it takes 365 days to complete one orbit while Mars takes 687 days. So, sometimes the two planets are on opposite sides of the sun, and very far apart. On other occasions, the earth catches up with its neighbour and passes close to it. It is this decreased distance during what is called the 'closest approach' that allows fuel-conserving space flights to Mars every twenty-six months, along with a glimpse of the Red Planet once or twice every fifteen or seventeen years. India's Mars mission was based on this earth-Mars-sun geometry.

Space Vision India 2025 outlined the need for planetary exploration to understand the solar system better. The sun, moon and Mars were the main focus of this concept. When the Advisory Committee for Space Science zeroed in on Mars, ISRO began putting together an inter-disciplinary feasibility study in August 2010.

ISRO's aim was to successfully enter the Martian orbit when its Orbiter spacecraft was only 500 km away from Mars, using the minimum energy transfer orbit approach, with the least amount of fuel consumption. A precise knowledge of the planetary movements around the sun made such a mission possible. November 2013 was selected as the best time, with the next possible dates being January 2016 or May 2018.

A 'simple' description of the Mars mission provided by scientists who worked on it goes like this—'It is like shooting from a moving platform (earth, moving at a speed of 107,000 km per hour around the sun) at a moving target (Mars, moving at a speed of 86,870 km per hour), the distance between them changing with time.'[19]

For the Orbiter spacecraft to follow a precise path with specific velocities at specific times throughout the

300-day journey meant making sure that the rocket which launched it into space, the onboard autonomous systems, the five payloads[20] and the ground-based tracking and telemetry—all needed to work perfectly in sync—to culminate in that moment of insertion (MOI) into the Martian orbit on 24 September 2014.

The 1,337 kg spacecraft was thus loaded with 850 kg of fuel and 15 kg payloads to detect, among other things, the presence of methane in the Martian atmosphere.[21] This mission, in addition to other scientific studies, would help determine the nature of past, present or future habitation on Mars.

The autorickshaw-sized satellite sent out of earth's orbit on ISRO's most reliable rocket, the PSLV-C25, on 5 November 2013 from the first launch pad at the Satish Dhawan Space Centre (SDSC), Sriharikota Range (SHAR), had to complete a 667 million-km journey in 300 days. The launch cost was an estimated Rs 454 crores (US$74 million).[22] This was 8.8 per cent of ISRO's total annual budget of Rs 5,172 crores for 2013–14.[23] Each Indian paid less than Rs 4 towards the Mars mission.[24]

In comparison, a similar NASA mission cost US$671 million, while a Hollywood film like *The Martian* cost US$108 million.[25]

The Team

From the feasibility study's institution in August 2010 to Prime Minister Manmohan Singh's announcement of the Mars mission on 15 August 2012 and the actual preparations, the MOM journey took 15–18 months.

Those Magnificent Women and Their Flying Machines

A team of 500 scientists from ten ISRO centres were roped in, with at least 27 per cent of the key executive positions being held by women.[26] Nandini Harinath and Ritu Karidhal handled mission operations at the U.R. Rao Satellite Centre (URSC), Bengaluru, in addition to calculating the spacecraft's trajectory to Mars and designing an autonomous software system to self-correct problems, respectively. Moumita Dutta and Minal Sampat built and tested the scientific instruments at the Space Application Centre (SAC). Several other women scientists also played significant roles in MOM.

The Wikipedia page on women scientists in India claims, 'Women scientists are fewer in number than men, they occupy fewer positions of power and face distinct issues by virtue of their gender and the accompanying societal pressures. Women scientists in India also tend to be less visible than their male counterparts, and public awareness of Indian women scientists is low'.[27]

ISRO's women scientists, particularly those that worked on MOM (or Mangalyaan), are slowly yet emphatically changing this paradigm through their performances, perspectives and plaudits. Nandini Harinath says laughingly, 'Many times I wonder about all the attention that we women scientists are getting... my male colleagues who work equally hard also deserve it, right? But sometimes I also feel that women need those kind of examples, so that they know it [success in science] is possible and they should not give up.'

Systems Engineer Minal Sampat is refreshingly frank about being in the spotlight: 'Recently a female colleague remarked, "you all are only working because you have been showcased." I told her it's not that we seek attention and go and tell the media about our efforts. We may be the face sometimes, but we are representing all of the women here.'

For the first time in ISRO's fifty-year history, a space mission charted a path markedly different from the organization's customary, low-key profile. The MOM invited people to participate in its journey of discovery. MOM project director, the tall and amiable S. Arunan, explains the rationale behind this strategy: 'With normal ISRO programmes, we don't go to websites and announce the kind of mission we plan to launch or the special features of the satellite. But for the first time, we especially wanted to involve and inspire our younger generation.'

This kind of practice is routine at NASA and the European Space Agency (ESA), where pamphlets, quizzes and puzzles are distributed in schools and colleges at the start of a programme to ignite curiosity. 'I believe this is one of the reasons they [foreign space agencies] are very good in that kind of a technology race. So we thought of doing the citizen's part of the mission. We got permission from the Prime Minister's Office to open Facebook and Twitter accounts and publish more information from the start of the mission,' confides Arunan. (ISRO employees are otherwise prohibited from maintaining social media profiles or using social media tools within government premises.)

This unprecedented approach led to surprising results, confirming Arunan's faith in the next generation of aspiring scientists. 'When we did our first manoeuvre—

our manoeuvres are done mostly around midnight or later—we published it on our website and on Facebook. We mentioned our calculations—"Our engine will be fired for this particular duration at this much velocity and at the end we will get this kind of an orbit." We posted that first manoeuvre around six or seven pm, and by ten pm the responses were pouring in. Messages like "good", "superb" were the most common. But there were also messages from young students in the age group of twenty to thirty years, saying, "Sir, you have mentioned one velocity—1.2 metres per second, but we calculated and it is 1.23, can you please check?" I was really enthralled to know young people were interested enough to do advanced calculations. So we set up a feedback mechanism—giving answers, explaining what we did. I felt that my faith in the student community was validated. You get feedback only when you share—till then you don't know what lies dormant in society. The confidence and support we got by including the younger generation of India in our journey was tremendous.'

MOM continued to share every step of the Mars journey with the public on Facebook and Twitter, with the 'likes' and compliments pouring in at every stage. This made the Mars journey almost a collaboration with the public. 'Earlier, ISRO would launch a satellite and inform people when it reached its destination, followed by a news item in the papers—sometimes small, sometimes big—depending on the nature of the satellite, or if something new was being done. Otherwise, it was business as usual,' says Arunan.

The MOM was different than the earlier missions.

Women from Mars

Apart from going public on social media, the other striking feature of this mission was that it allowed the country to see women scientists upfront for the first time, dispelling long-held stereotypes of the fuzzy-haired Einsteinian male scientist. 'You know, we know Mars is for men. Now we have proved that Mars is not *only* for men,' Arunan jokes. 'When we formulated a young team for MOM, the time given to us was just eighteen months. We needed the bubbling energy of the younger generation. We had a lot of problems that needed to be solved to complete the mission successfully. The women's team had worked on Chandrayaan-1 [ISRO's mission to the moon in October 2008] and other challenging earth missions. MOM was a calculation-intensive mission. The navigation or mission aspects called for iteration [repetition of a mathematical procedure] calculation. Some of the women are very good in pure mathematics. Through all of our manoeuvres, designs and calculations, the women's team stood by their decisions. Their accuracies were fabulous. I did not see them slipping even once in their calculations. In reviews [there were multiple reviews during MOM's eighteen-month duration] you have to be able to justify your decisions and calculations. You need to be convincing for the reviewing committee. This particular team, including Nandini and Ritu, was steadfast,' Arunan claims.

Nandini Harinath remembers the suspense and the excitement at the time of the MOM team selection. She wondered whether it would be sanctioned at all—the costs involved were high and the project had no direct social application, unlike other ISRO satellites—and if the study team would translate into the project team. 'The Prime Minister's speech on 15 August confirmed that the project had been sanctioned. And then the rumours began—who would be the project director? Who would be part of the team? As speculation floated around—we are a big team of 3,000 people at the URSC—you keep hoping, but till you actually know, you don't really want to believe it,' Nandini laughs.

Nandini was in the thick of another mission at the time—as operations director for RISAT-1 (Radar Imaging Satellite), India's first microwave imaging satellite at ISTRAC—when she was asked to return to URSC to be one of the 'focal points' of the Mars mission. 'We weren't designated, we were just told to start working on it from Day One. "Find out what others have done and do what you're supposed to do"—those were the orders. And that started the mad race right from the first hour,' recalls Nandini. She quickly wound up her pending work and began downloading information and studying the data available.

'We were a big team, with two deputy operations directors instead of one, Ritu and I, looking after different systems. We would have frequent meetings about what each of us had found. Most of us in the team were first-timers for this kind of a project.' They had approximately fifteen months, from start to finish, to build an autonomous satellite with no prior knowhow.

Ritu Karidhal had finished her work on the launch of RISAT-1, when she was called by her group director, V. Kesava Raju, and given a deadline to get the Mars spacecraft ready so that the final mission could be accomplished in eighteen months.

Women from Mars

Ritu elaborates on the formidable task they faced in 2012: 'I had worked on Chandrayaan-1 but not as part of the core team. In any case, Mars was a totally different project where we had to go beyond the earth's sphere of influence and gravity. It was an interplanetary mission, so we were very excited.'

'The excitement over the Mars mission was natural,' says Arunan. In a country like India, which was using different techniques for the first time (in Asia) where others had failed, the whole nation was enthusiastic in its support. 'Moreover, Mars is seen as some kind of possible colonization later on, so everybody is more interested in it than in Saturn or Jupiter or other missions.'

MOM Project Manager Moumita Dutta agrees: 'This mission triggered immense curiosity. Everywhere I go the first question asked is, "Does life still exist on Mars?" followed immediately by, "In the film *The Martian* we saw people going to Mars. So can we think of going and living there some day? Is it possible?" I say that nothing is impossible.'

'Everything was new to us. As mission designers we were working for the main core or what we call "bus" system of the spacecraft which was to travel from earth to Mars. [Once the craft started revolving around Mars, the payloads designed by SAC scientists like Moumita and Minal would come into play]. For us the challenge lay in managing the entire mission itself. Moreover, since any control or command would take a minimum of twelve to twenty-four to forty-two minutes to reach the other end, the spacecraft had to be very, very smart and autonomous,' says Ritu.

Those Magnificent Women and Their Flying Machines

Arunan gives an example of the out-of-the-box thinking that led to some innovative solutions: 'We had a launch vehicle that didn't give us sufficient energy for a direct path to Mars, compared to foreign countries that have high energy vehicles. That meant we had to think slightly differently than others—we had to come up with solutions that would add energy without spending more money on fuel. One of the ideas was to use the moon's gravity to impart energy to the spacecraft. Gravitational forces of other planets were made use of without us doing anything—this was a great example of lateral thinking by this particular navigation and mission team of women.

'Contingency plans were another area where the women excelled, in the precision of the calculations involved, which were in centimeters and millimeters. This optimization was done in a very economical manner by this team.'

I ask about the roles of the women in the project, if they were on an equal footing with their male counterparts, and if the latter were supportive.

'The women played integral roles, not subordinate to anyone. They were team leaders in their own right and expertise, assigned 100 per cent responsibility and accountability. And they delivered with the precision demanded by the mission. There were, of course, many women who played supportive roles in MOM. But in the major areas of the mission—navigation, communication, control systems, spacecraft design and tracking—the women were leaders,' Arunan responds.

In an organization where women scientists make up

less than 20 per cent of the total scientific and technical workforce, the Mars mission saw women shouldering a fair share of the responsibility, according to Arunan. Tapan Misra, Director, SAC, says, 'Women are the backbone of the organization—they do the work yet remain invisible.'

If we uncover these 'hidden figures', what are we likely to discover? How hard was it for these women scientists to follow their dream of a career in science, which has been considered a strictly male bastion? How many women made this choice of their own free will? How many were hampered by family pressures and social conditioning positing marriage and family as their primary goals? How many of these women encountered the glass ceiling in the corridors of power?

Did the demands of family distract them from their focus? Did they feel torn or guilty while prioritizing work over family, especially their children? What are their coping strategies for dealing with stress and exhaustion? Are these scientists rewarded for their achievements equally or are promotions and recognition still given predominantly to their male counterparts?

In the following pages, the women scientists at ISRO—the Mangalyaan team members as well as those in other disciplines like navigation, remote sensing, communication, applications, space science—provide answers to these questions, along with startling insights into their fascinating yet lesser-known worlds.

These scientists are visible, shining role models for

millions of young girls across India who want to script their own story of change, navigate their own odyssey into space and explore the mysteries of science. Their time has come and hopefully, for this generation, the sky will not be the limit.

Chapter 2

MOM: Operations in Outer Space

I think the human race has no future if it doesn't go into space.
—Stephen Hawking

'Into Outer Space, Launching Mangalyaan, Mars Orbiter Mission—Ritu Karidhal, Moumita Dutta, Nandini Harinath, Minal Sampat, Indian Space Research Organisation,' reads the invitation sent out by the Indian Women Network-CII Maharashtra for 'WomeNation: Power of Us', a two-day summit on Indian women achievers in September 2016. 'India's rocket women' are scheduled for a 45-minute session, longer than the 'Women in Media', 'Durga in the Boardroom' and 'Lunch and Networking' sessions scheduled around it. It is an instant draw for me.

I have only a hazy recollection of what these women scientists actually did during the Mars Mission two years ago, and am eager to know a lot more. I enter the large hall with several neatly arranged round tables with spotless white tablecloths, crystal glasses and water jugs. The audience comprises mostly of women from the corporate sector, who have all paid to attend this celebration of women professionals in diverse fields. Some appear to be senior management leaders busily giving muted instructions to others at their tables, while the majority are young, stylish professionals at junior and middle-management levels, along with a smattering of entrepreneurs and students.

Not all of them are paying close attention to the speakers onstage: the room is humming with the usual low-pitched conversation and distracted texting. Polite applause greets the end of the mid-morning session.

The session moderator introduces the women I have battled two hours of stressful Mumbai traffic to hear—Ritu Karidhal, Minal Sampat and Moumita Dutta, the Mangalyaan scientists 'who happen to be women'. Nandini Harinath could not make it due to work assignments elsewhere.

For the next hour, as these three women held the audience and the organizers equally spellbound, an idea slowly gathered shape in my head. With slides and statistics, drawing word pictures and vivid diagrams, the ISRO scientists took the largely non-scientific crowd through a journey into the cosmos. They spoke of incredible tasks completed within impossible deadlines, juggling between family and work, the diverse challenges and moments of accomplishment. It was an all-too-brief glimpse into the realm of the unknown, of inter-planetary travel, of possibilities and promise.

Soon, the impending lunch break cut short the proceedings, leaving little time for questions from the obviously impressed audience. A few did manage to ask the scientists about their work schedules, the pressures, their coping strategies and Minal, Moumita and Ritu took turns to answer.

Within sixty minutes, overshooting their allotted

time, they clearly owned the room. Though the entire interaction was clearly aspirational for the women-centric audience, one phrase in particular stayed with me: 'We worked on the mission during the day, we often worked nights as well, and in between we looked after our children and families'. This, a throwaway statement by one of the scientists, was accompanied with laughter and acquiescence from the others.

This matter-of-fact attitude, the underlying confidence shining through the 'balancing act' they professed to perform on a daily basis, is perhaps the most striking characteristic of the ISRO women scientists. They reminded me of the iconic Tata Steel ad of the early 1990s—'We also make steel'.

I decided to skip the post-lunch sessions devoted to the other women achievers and went looking for the Mangalyaan scientists. They were preparing to give pre-arranged interviews to a women's TV channel. Before the interviews began, I asked them if I could meet them for another interview—at much greater length—to write a book about their accomplishments. All three looked visibly pleased but told me they were not allowed to grant interviews without prior clearance from their organization. I would need to get in touch with the public relations director at ISRO headquarters in Bengaluru. With my permissions secured, I started my own mission over the next few months, to get to know these magnificent women and their flying machines.

Ritu Karidhal

Deputy Operations Director, MOM

Four months later, I sit patiently at the reception desk at URSC in Bengaluru, waiting for the stamped papers that would allow me to carry my Dictaphone and mobile phone inside the premises, so I could record my interview with the Deputy Operations Director, MOM, Ritu Karidhal.

The URSC complex has a sprawling old-world charm, incongruous to the elaborate security measures. An impressive canopy of greenery—with ancient, gnarled trees and several rows of recently planted additions—liven up the otherwise staid government structures on campus.

Armed with the required pass, I enter the main building where my bags and person are checked again, before being politely waved through to the first floor.

I wait for Ritu in the airy, well-appointed room and she enters exactly on time in an elegant salwar kameez, which prompts compliments from the secretary who follows with a coffee tray.

Ritu has a pleasant face and a slightly stocky build. She wears rimless spectacles and a red bindi. She has a quiet, earnest air about her and an unhurried way of talking—especially while explaining complex scientific procedures.

As she speaks eloquently about 'slingshot manoeuvres' and the 'velocity near the perigee and apogee', I begin to wish that I had paid more attention in my school physics class. Noticing my confusion, she patiently gives me a crash course on the basics of how the Mars Orbiter satellite operates.

Then we backtrack a little to her childhood dreams and aspirations. A flashback reveals an uncommon fascination with the universe, particularly the moon. 'When I was quite little, maybe three or four years old, I remember we used to travel by those hand-pulled rickshaws in my hometown, Lucknow. And I would think the moon is walking with me wherever I go. When I slept on the terrace, I would feel the moon following me everywhere. So I'd wonder about it, and ask myself these questions even before I went to school,' recalls Ritu.

I ask her what sparked this curiosity so early in her life.

'I was a calm and quiet child, who didn't talk much and preferred to read. Even in the extreme summers in Lucknow, I remember I would get lost in my book and forget to switch on the fan. And then my mother would come in, scolding me and asking me to rest,' says Ritu, smiling nostalgically.

At school Ritu was a good student, excelling at mathematics. She also had a vivid imagination. 'I would picture myself surrounded by numbers, writing poems related to mathematics,' she chuckles. School was fun, with supportive teachers encouraging her enthusiasm for science, along with her parents.

'Twenty years ago there weren't too many opportunities for girls in small towns. We didn't have a lot of facilities. Even going to college from my home, I had to take a rickshaw or auto or bus. Travelling wasn't easy but I had that inner passion, that curiosity to know and to do something. That perseverance convinced my parents to support my studies. In so many families, girls cannot do this. They are only expected to work at home. This kind of thinking decelerates their growth.'

After a six-month stint teaching physics at Lucknow University, Ritu was selected for her very first job at URSC, ISRO, in Bengaluru. She had always kept a meticulous track of what ISRO and NASA had been doing, reading and preserving newspaper clippings about landmark events, such as the first NASA rover landing on Mars in July 1997. She had also looked out for ISRO's advertisements for vacancies, and when one did come, she quickly applied. Her graduation and post-graduation grades in physics (later followed by an MTech in aerospace engineering from IISc, Bengaluru) and her performance at the ISRO interview led to an immediate selection and a dilemma for her parents.

'Lucknow to Bengaluru is a distance of more than 2,000 km, so my father took one week to decide if I should join ISRO. My mother assured him that I would be able to manage, and I came here,' says Ritu. She is also thankful that her parents didn't pressure her about marriage. (After a few years she got married in the traditional 'arranged' way to Avinash Karidhal, an engineer working with Titan).

Ritu joined ISRO as a young engineer in November 1997. Her physics background led to her first posting in the Mission Analysis division with Dr Kesava Raju (who later became the Mission Director for MOM) as her boss. 'The first problem he gave me was very difficult—how to manoeuvre a satellite to get a stereo image from the onboard camera. I was very happy that I could straightaway apply what I had studied in physics and mathematics in my first job. I completed the work he gave me in three or four months, and it was implemented onboard. Later, I got more projects, which were all very challenging. People talk about MOM deadlines today, but even then there used to be projects where we were very hard-pressed for time and the targets had to be met,' she says.

The deadlines were often as short as three or four months, and Ritu remembers working round the clock to meet them. 'I was a bachelor then and lived as a paying guest. I would work continuously from morning to evening, but I enjoyed it because I liked my work. Not many girls were working at ISRO at that time and when I stayed back at the labs, there would be very few people around. Even the walk down the long corridors from one lab to another, or from one building to the other in the URSC campus would be fairly lonely. But I never felt any fear.'

The universe was, and continues to be, her workspace. Besides her own passion, Ritu credits the senior scientists who expressed confidence and trust in her capabilities for her success.

I ask her about the work environment at ISRO, and more importantly, the gender equation, given that today women constitute approximately 18.8 per cent of the total workforce—3,188 out of 16,902 employees—with an even lower strength of 16 per cent in the technical category (1,978 women out of a total 12,300 scientists and engineers).

Those Magnificent Women and Their Flying Machines

In keeping with the overall position of women scientists across India, where scarce or almost no updated data is available for those holding top positions, ISRO too has not had a woman chairperson or director for any of its sixteen centres. Although several women are project managers and project directors, while a smaller number are senior programme directors, group directors and deputy directors.

'ISRO provides a very positive atmosphere. What matters here is your talent, not your gender. You get challenging work. For my first assignment, many senior men were eligible, but it was given to me.' She concedes that the stereotypes in the world of science—the image of the male doctor and the female nurse, for example—need to change. 'And it *is* slowly changing,' says Ritu, with ISRO itself breaking the glass ceiling gradually. 'Senior women scientists in the fields of remote sensing and communication satellites have become programme directors, and once the numbers increase, a woman director will not be a rarity,' she adds. 'Nor will the public recognition and awards for women scientists be as novel a phenomenon as it is today.'

India's top national science award, the Shanti Swarup Bhatnagar prize has been given to over 500 scientists since its inception in 1958. Only sixteen of these have been women.[28] In 2017, not a single woman took home an award. Not enough women are nominated for these awards to make the shortlist.

Globally too, out of all the 881 Nobel Prize winners from 1901–2016, only forty-eight were women, while the Fields medal for mathematics has only once had a female winner—Iranian-born Maryam Mirzakhani in 2014. Ritu received ISRO's Young Scientist award in 2007 for her work in mission planning and operations.

She pays little heed to unspoken discrimination, whether it is a condescending remark by male colleagues or the fact that women are often overlooked for promotions and patronage at conferences and committees. 'I don't worry about what others might say. People can say anything, it does not matter to me. Even at home you find relatives or friends commenting in a certain manner, but ultimately when they see the depth of the work or its relevance to society, or view ISRO as a whole, they change their views.

'God has created everyone equal, but due to pigeonholing—"Your work is this and only this"—they are not able to break out of their assigned gender roles. But if you work without the fear or assumption that others will stop you, or that you won't be able to manage so much, if you break the internal glass ceiling first, those roles can change. The rules can change. You need to first show how high you can climb. Someone else not allowing you to get ahead—that comes later,' she avers.

Ritu's mantras combine pragmatism with hope, as she outlines how they have worked well in her own career. 'Family, marriage, pregnancy breaks, children—these are all part of life and cannot be treated as mutually exclusive from your work,' she reiterates. Time management is the key to a successful work-life-family balance for Ritu—a refrain I will hear several times over at ISRO.

'When work began on MOM in 2012, my son was

nine and my daughter was four, but it wasn't as if I was only working and not available at home. Because if I am a proper mother, my own feelings will not allow me to do that. So we have to multitask and double the effort. Work here, then go home to your family and spend time with them... Then start your work again late at night when they are asleep. I did many things from home.'

Ritu configured a small setup on her laptop, and used to be up till three or four am, completing tasks for MOM. 'I did feel physically exhausted, but you can overcome this exhaustion in different ways. When you see the output and what you've achieved by putting in extra effort, that is worth it,' she smiles.

Did she have to work twice as hard as a male colleague to be taken seriously in the office?

'Not really. We put in as much effort as our male counterparts do here, but then we have to put in equal effort once we go home. So we are a bit hard-pressed for time. But I feel women are capable of doing that... There were times when it was difficult to manage both home and office. I remember a time when my daughter had high fever and I was not able to take her to the doctor. My husband took care of her. I called regularly to ask about her fever—I felt really guilty about not being with her. There were times when I couldn't attend school functions or PTA meetings or be home when I was needed. But the one thing that gave me strength was my family's support. I am convinced that without the help of family and friends, it is not possible—for a woman at least—to cope with the increasing demands of the workplace.'

Does she think that the traditional mindset will change, so that the family is equally the man's responsibility?

'They [husbands] may not be able to fully do what we can do at home, but if they support us at the time of need, if they understand and at least allow, that itself is huge. Some women would not be allowed to stay back late in office, but my family understood the high pressure I was under. Mars, especially, was something new for all of us... I would come home late many times, work odd hours, but my children rarely complained and my husband's understanding made a big difference. If I had had to face problems at home, then my mind would not have been at peace and my work would have suffered.'

~

The lift isn't working in the building tucked away inside a small lane in a bustling residential complex, fairly close to URSC. I trudge up four floors, panting for breath, and ring the doorbell. A young boy, about thirteen years old, answers the door, introducing himself as Aditya—Ritu's son. His parents aren't home yet. He offers me a glass of water, engaging me in polite conversation till Ritu appears, looking visibly tired after a long day.

A little girl with tousled hair peeks out from the doorway, feeling shy for all of five minutes. Thereafter, 8-year-old Anisha holds forth, once I get Ritu's permission to talk to her children. 'I give Mama water when she returns from office. I talk to her and then play with my friends till seven pm,' she says importantly, proceeding to tell me the names of all her school and building friends.

'Mama comes home very late sometimes. She works till midnight, sleeps at one and wakes me up at six-thirty in the morning for school. She works very hard and is a space scientist. I am proud of Mama and want to be like her. I tell all my friends about her.'

She goes on to spill the details about the television serials Ritu likes to watch, the films they see, the holidays they go for, or don't. Referring to her brother, she says, 'Bhaiya used to complain during exam time—' but Aditya cuts in with an earnest defence of his mother. 'It was Ma who helped me with my exam preparations even during MOM. Sometimes she would sleep at the URSC guest house. I am proud of my mother. I want her to work, to promote women. I want India to be the best in space.'

Aditya attends ISRO's science exhibitions and visits planetariums. 'I like to find out if there is something new discovered in space,' he tells me solemnly. He talks knowledgeably about the 'onboard autonomy' of the Mars satellite, and how his mother put in long hours for the mission. Insisting he is now 'all grown-up,' he describes how he was taught to be independent by his mother, and to study on his own using Byju's smartphone app. 'I got 98 per cent in science,' Anisha butts in, 'but I am scared to be an astronaut because they can die like Kalpana Chawla.' The children have an independent routine on school days—a wake-up call at six am, school from eight-thirty am to three pm, a couple of hours at the creche and then another two hours or so at home till their parents return around seven-thirty pm for an early dinner. Weekends are reserved for the cricket academy for Aditya and films for the rest of the family. They have recently seen Sachin Tendulkar's biopic, *A Billion Dreams*. 'Bhaiya wants to be a cricketer,' chirps Anisha. 'I want to be like Mama, do hard work and never take shortcuts,' retorts her brother sternly.

The tea break following the chat with the children is extended as Ritu's husband, Avinash, is stuck in traffic. Their home is modestly furnished. The drawing room has old-fashioned sofa sets adjacent to newer, modern ones, a glass table and a large wooden cabinet with several medals and photographs. Prominent among them are framed pictures of Ritu receiving the Young Scientist Award from President Abdul Kalam, another with Prime Minister Manmohan Singh in 2010 and other felicitations from state governments. A few sports medals won by Aditya occupy the pride of place. A MOM-replica clock keeps time on the wall.

Avinash enters in a flurry of apologies, proceeding to shower me with snacks and more tea, followed by an invitation to join the family for dinner. It is the kind of warm hospitality rarely seen today. A few minutes into the interview and his pride in his wife is evident, heartfelt and openly articulated.

'Can we say in your case that behind a very successful woman lies a man who has supported her?' I ask.

'Our family support is always there indirectly, but ultimately it is Ritu's effort and dedication... In fact, today I would say that my identity is through my wife. She meets many top dignitaries like the prime minister, president, state chief ministers and MLAs. Recently, she was honoured in Lucknow and she got an award from the chief minister of Kerala. So I am now known as Ritu's husband,' Avinash says. 'That identity is growing and I'm happy,' he adds with a laugh. 'Not just because she is my wife, but because I believe that those who are in such fields of work—research or science and technology—should progress with full dedication and honesty. Let them rise and get a good name. In our extended families, both hers and mine,' he points out, 'our relatives derive their identity from Ritu and her achievements. Whether it's the father-in-law or brother, people know them because of her.' In Allahabad and Lucknow, Avinash and Ritu's hometowns, respectively, their in-laws and parents constantly boast about Ritu's accomplishments and public appearances. 'The folks in Allahabad and Lucknow take more pride in her achievements than her family back in Bengaluru,' jokes Avinash.

An engineer working as a supply chain manager with Titan industries for the last twenty years, Avinash always knew the importance of work for his wife—right from the time of their marriage proposal. He was 'well prepared for it'. He seems slightly in awe of his wife's job, calling it a service to the nation, in contrast to the more commercial nature of his own job, 'where one good promotion means a salary hike of 20–30 per cent, or becoming a general manager or vice president. But it isn't nation building.'

What would he say if his wife were to aim to be the first woman chairperson of ISRO?

'Whatever she wants to achieve, let her achieve. Let her climb as much as she can.'

MOM: Operations in Outer Space

'He supports me a lot. Sometimes we also have to not be too demanding over some issues. You need to balance, otherwise just making headlines will not help. So there are compromises on both sides,' Ritu comments.

How often are the roles reversed or shared, as far as looking after the household and children is concerned?

Avinash claims that he does 90 per cent of the work. 'She doesn't even know if there is rice and dal in the house or not,' he laughs good-naturedly. She takes care of the children's studies and he looks after all the payments—electricity, society maintenance, filing Income Tax returns and other sundry tasks. However, 'Coordinating our leave is quite a task—we have missed out on so many vacations because she is not free when I am, and vice versa. But we try to manage.'

The fifteen-odd months of the Mars mission were quite challenging, agree both husband and wife. 'The children would miss her a lot, because she often went to ISRO at midnight. Some kinds of operations would take place all night. When launch operations happened, there would be no time at all—no holidays or weekends, nothing. If there was any issue with the satellite or any kind of correction to be done, she would rush for follow-ups,' recalls Avinash.

'Ritu also travelled extensively. She would go to Delhi for three or four days, Thiruvananthapuram and Ahmedabad. She would take early morning flights at five-thirty am, so I would be awake at four am, get the children ready from six-thirty onwards. We have helping hands, of course, with a full-time cook and a driver.

But the responsibility was mine.' Managing his own work pressures and keeping the children on track with their studies and school did clash sometimes, 'but we ultimately worked it out,' he says. 'I often worked on my laptop from home,' Ritu reminds him. The hardships all paid off in the end when Avinash and his family watched MOM's—and Ritu's—journey on TV.

At the Titan Bengaluru office, all of Avinash's colleagues and bosses know what Ritu does, often alerting him to sightings of her photographs or profiles in newspapers and magazines. He himself stays abreast of his wife's work—the kinds of launches she is involved in, the basic operations of satellites and the commercial applications such as geographic mapping or tsunami alerts. He predicts that with additional launches, private partnerships and technology tie-ups for satellite manufacturing, life at ISRO is bound to change markedly.

Would it not get tougher for the family?

'Yes, of course, but what we both feel is that as long as we are healthy with good brains, we should inspire others and not sit idle at home,' says Avinash as Ritu agrees wholeheartedly. He adds with a smile: 'Every two months there is some satellite triumph happening, so it has become kind of routine for us.'

Avinash's comments reflect the general esteem and respect that ISRO and its scientists—as architects of a truly indigenous success story—command, of late, from the general public as well as those in positions of power, across political hues. But till the Mars mission began, the average Indian would not have been able to recognize any of the scientists who dreamed and worked towards placing a satellite millions of miles away in space.

~

'How is it that you people are not as famous as film stars and sports icons?' I ask Ritu, back at her office. 'Your work is so exciting, don't people want to know more about it and about you?'

'People will want to know about us and our work only if they are exposed to it. It's not like we can go and project ourselves to the public. The media or those who can reach out to people, to the masses, if they take initiatives to popularize the work, it can happen.

'The people are interested,' she continues, 'but platforms are required. We have so many film magazines but no science magazine. Maybe we should also have a Sciencefare magazine, in addition to *Filmfare*... Everyone wants to see entertainment, not knowledge-based programmes, though gradually now the trend is changing with documentaries and articles on science.'

Ritu gets little downtime. On a break, she is generally found meditating or sitting quietly and praying, as part of her own relaxation therapy. She believes that thirty or forty minutes of quiet contemplation gives her the energy needed for the day's work. Unwinding with her children is also a relaxing pastime. Despite a heavy workload at URSC, Ritu takes time out for TED talks and interactions with schoolchildren. 'They ask a lot of questions and get really animated, listening to what I have to say,' she beams.

'Maybe one of them will ask you why the moon follows *them* around,' I chuckle.

'Yes, children just need the right kind of exposure. We have to increase our outreach in rural India too, there are so many places to go to.'

~

Ritu's 21-minute TED talk on the Mars mission in Hyderabad in June 2016 is a masterclass in its comprehensive and concise chronicling of the Indian space story. It is peppered with slides, diagrams and anecdotal references to Mangalyaan's trajectory, right from its inception. And as the narrative is kept intentionally simple and free of scientific jargon, the audience reacts with rousing cheers at regular intervals.

The biggest applause is reserved for the list of 'India-first' statements—India is among the top six space agencies in the world; Mangalyaan is the most economical project ever planned in the interplanetary arena; the first with an impossibly short timeline and the first Indian satellite with full-scale onboard autonomy.

She explains that 'onboard autonomy' means the satellite is capable of self-diagnosis and self-recovery, and can execute all the instructions loaded onto it in advance, with high precision. This is crucial, since there can be a 24–42-minute interval between the sending and receiving of a message to interplanetary satellites like MOM. The minutest error in diagnosis would lead to failure of the mission.

Another challenge was the tracking of the last stage of the PSLV launch rocket, which was to eject the spacecraft into space over the Pacific sea in Australia, where no ISRO ground stations or Deep Space Network antennae were available. Two ship-borne terminals were thus deployed and took two months to reach the location. Bad weather conditions prevented the ships from reaching on time. They finally arrived in place just before the window closed. 'The launch happened with the mighty PSLV-C25 taking off on 5 November 2013, when the satellite was sent to its first parking orbit,' Ritu proclaims to resounding cheers from the audience.

Using diagrams, she shows how with each burn or engine firing operation, the satellite's orbit gets bigger and bigger the more it moves away from the earth, till after twenty-five days, it has to come out of the gravity well of the earth and move towards Mars. This is known as the cruise phase. The trajectory has to be 99.99 per cent accurate. 'Many satellites fail while leaving the earth's gravity but ISRO succeeded and made history on 30 November 2013, when it put the satellite on the correct heliocentric or sun-centric path,' says Ritu, to loud applause yet again.

~

While the TED talk lasts for less than half an hour, Ritu is a lot more relaxed and forthcoming at her workplace at URSC, sharing many personal stories of her MOM journey with me. It is a story that bears retelling, because listening to different perspectives helps one understand the magnitude of the challenge and the odds of its unlikely success in the very first attempt. As Ritu explains, the scientists working on MOM had to anticipate and overcome hurdles on various fronts.

'Mars itself was driving our deadlines,' laughs Ritu,

'because if we hadn't been able to launch when we did, we would have had to wait for another twenty-six months.' Once the launch date was fixed, the mission operations team had approximately ten months to conceptualize, prepare and execute a plan that combined accuracy and ambition, in equal measure.

'Designing a 99.99 per cent accurate plan in ten months… I'm getting tense just thinking about it,' I tell Ritu, and she smiles.

'We had discussions with engineers and scientists, and we debated the changes for each and every possibility. We had run tests for all possible cases and even made it fail to see if it survived failure.' All of the testing was done at the URSC, with every level tested by different teams.[29]

Later in the day I am escorted to the glass-paned viewing gallery at URSC to observe the 'clean room' from above. A satellite had just been dispatched to be integrated into the rocket launch vehicle at the Satish Dhawan Space Centre (SDSC) at Sriharikota. The humongous size of the room below—about 55x34 m with a height of 60 m—is astounding. Technicians and engineers in gowns and protective caps work busily, often in simulated conditions similar to space.

Ritu continues, 'When the spacecraft leaves the earth's gravity, it has to leave at a particular time in a particular direction, and it will get that direction and escape velocity if energy is supplied correctly by the engines and the propulsion system. The flight path angle is very important because if you miss, then instead of going to Mars, it will go somewhere else. You have only one opportunity, a second attempt would mean wasting more fuel, and there would be penalties for that... You need a perfect 99.99 per cent performance. For that you need to work backwards—imagine all that is required and then make it happen. So it was a tough task.'

NASA's Maven Orbiter had a different plan to enter Mars' orbit. Since they had no weight or fuel restrictions, their spacecraft could take a direct path. ISRO's scientists tried out a new concept of slingshot manoeuvres. Ritu explains, 'Every time you come near the earth and fire your engine, you get more energy. So, utilizing that energy takes you very far. Again, you come back to earth and fire, and again the energy you get takes you a little further, because the earth's gravity itself provides you more velocity near the perigee [closest point] than the apogee [farthest point]. So if you fire at that point, you increase your altitude with less consumption of fuel. Each time the size of the orbit becomes bigger, the velocity also increases. This is termed as the "multiple slingshot theory". You also have to do all this at predetermined dates and times.'

Six orbit raising manoeuvres [precise firing of the liquid rocket engine] took place from 7–16 November 2013, till the Trans Mars Injection (TMI)—the crucial slingshot to give a boost velocity of 650 metres per second—set the spacecraft onto a sun-centric trajectory on 1 December. It would reach near Mars on 24 September 2014.

'Such high-pressure work,' I marvel, and Ritu immediately clarifies.

'People know us because we are the submission

people, but there is a large, fully dedicated team behind us working together—thermal, communication, power systems, controls, onboard computers and many more. Realizing a mission like this can never be a one-person task.'

'ISRO also designed this mission at one-tenth the cost of other missions to Mars, such as NASA's Maven Mars Orbiter, isn't it?' I ask.

'Yes that is now a record, I believe,' she replies, 'but it's not just the money, although it is important. I have seen the dedication and teamwork involved in this technological manoeuvre, to show the world that we can also accomplish such highly complex jobs as interplanetary travel.'

ISRO returned the leftover funds to the Government of India, another gesture that was appreciated by the Indian public. 'We now have various collaborators. NASA and other space agencies are approaching us to launch and lift-off their satellites from our PSLVs,' says Ritu proudly. The little girl who wondered why the moon became bigger or smaller is today a part of the Chandrayaan-2 mission, involving an unmanned rover landing on the moon in 2019. 'Manned and unmanned landings are totally different. In a manned landing, you are near the surface and you know where to manoeuvre the spacecraft and land in a safe place. In an unmanned landing like the one we are working on for Chandrayaan-2, you have to find the safe place with the onboard cameras. It has to be highly accurate and precise.

'Chandrayaan-1 was a very successful mission,' she

continues, 'with several payloads. A lot of science came out from that, such as the detection of water on the lunar surface. But Chandrayaan-2's unmanned landing will provide a technology which can eventually help explore habitation on another planet. It is the first step towards that,' she clarifies. 'We have to keep pace, otherwise we will be left behind.'

Her own trajectory, from walking alongside the moon on earth to actually going to the moon through the unmanned rover, has been the fulfillment of a dream.

'What dreams will you pursue now?' I prod Ritu.

'I want the work I am doing to have relevance for the common man. I should be able to do something which can improve his quality of life,' she replies. She rattles off the various space applications ISRO is involved with—disaster management, satellite data for fishermen about the best areas for fishing, telemedicine in villages across the country, tele-education, information for farmers about crops and weather, and much more.

Having proved that women scientists *can* go to Mars, are they now heading towards Venus?

'It may take some time but the Venus project is on the anvil. We also have MOM-2, with more payloads, so we can do more science. For MOM-2, we will be more relaxed.' Given the ISRO work culture of discussing, debating and once finalized, imposing impossible deadlines, relaxing may be easier said than done.

Nandini Harinath

Deputy Operations Director, MOM

It would not be unreasonable to assume that before the Mars Orbiter Mission, Nandini Harinath was known only amongst her inner circle of family, friends and colleagues. But now, when you search for her on Google, you get 2,11,000 results in less than 0.41 seconds. Nandini actually prefers to keep a low profile. She is uncomfortable being in the limelight.

However, I believe there *has not* been enough limelight on her and the other women scientists at ISRO, matching the quality of their work and its high-pressure milieu. ISRO is very careful about scientists talking to the media. Every public appearance, even those for educational lectures in schools and colleges, requires prior clearance from the management. No one can meet any ISRO scientist on their own. Fortunately, I have advance permission to interview the deputy director, operations, of the Mars Orbiter Mission.

~

On a sunny Saturday morning, Nandini drives into the spacious and tree-filled entrance of the URSC guesthouse in her compact i10, entering the special visitors' room where I await her. She has just returned from a two-week trip to NASA headquarters in the US, for a collaborative project called the NASA-ISRO Synthetic Aperture Radar (NISAR)—a radar that takes images at lower altitudes. She is severely jet-lagged. However, instead of taking much-needed rest, she enthusiastically participates in a free-flowing discussion about young girls and STEM, gender diversity, the need for more women icons, working at ISRO, and the Mars Orbiter Mission that made her famous.

'ISRO has always been doing very good work right from the beginning,' she says. 'It's just that the public has not been aware of its extent. None of this has happened suddenly, it has been built up over the years. ISRO also took the decision of going public with MOM. It got some very young people to update the public on social media on a daily basis, so many people understood what the mission was. That made it all very interesting.'

There is a whole lot of interest, I tell her, across a wide spectrum of the public: not just among the science-engineering-left-brained segment. Other people are drawn to this subject as well, especially if information related to such projects is relayed in simple terms.

Nandini laughs in sympathy at my non-science background and her eyes light up as she tries to explain the Mars mission as clearly as possible. 'When the project was conceived, I was designated as the project manager, design. Later, I was the deputy operations director like Ritu, basically doing the same work but with different systems in the satellite. I was doing things related to thermal power and communications. Communications at that distance was a real challenge.

'The satellite had to take decisions on its own, reboot itself, rig up the back-up system in case of any failure, since mistakes could not be rectified in real time by the scientists on the ground.'

Did she have to anticipate all the possibilities?

'Exactly. We had to identify all possible scenarios, work out all mitigation plans and then see how much of it could be put on board and how much could be made autonomous. This was the design aspect. Then when it came to operating the satellite, we were actually making the procedures, running simulations for all possible scenarios—if this happened or that happened how would we react and what would we do?

'These were all critical, one-time events—when the orbiter exited the earth or when it was to enter the Mars orbit, where the satellite *had* to work in the very first attempt... There were no second chances,' says Nandini. 'Meticulous testing followed, with ground simulators at URSC running simulations for all processes.'

Nandini was also handling the planning and execution of all activities of the payloads (scientific instruments onboard the Orbiter) within the constraints of the spacecraft. The lack of margin for error was compounded by the resource constraint. 'Meeting our goals within the given resources was the most challenging part of the mission.'

Multitasking and innovating within a limited timeframe along with financial constraints and tough targets—does this sound familiar? It could well be the textbook definition of woman power at home and at work. Only in this case, the arena is rocket science. Speaking at a seminar titled 'How to Deal with Sexism In Science' at the *India Today* Women Summit in September 2015, Nandini drew an analogy: 'The Mars Orbiter Mission exited the earth's sphere of influence to proceed towards its target orbit around Mars with a series of well-planned and executed manoeuvres. Similarly, for a woman to reach her destination, she needs to exit the stereotypical sphere of influence around her and design her trajectory to make that giant leap.'

Nandini goes on to tell me, 'My husband and I both work at ISRO. He realizes that when I am on a mission, I need to look after it. He steps in and takes over totally when I'm on tour or on an important mission. It's not that we always work 14–18-hour days. This is the case only when a project peaks or something demanding comes up.' Her husband, M. Harinath, works in the Control System Analysis and Testing section at URSC, and was part of the testing team for MOM.

Nandini discusses the need for a change in social perception. 'I am optimistic this will happen because in this generation, in fact even in the previous generation—my parents didn't differentiate between my brother and I—at least among the educated urban people, there is no problem at all. A lot of my colleagues have had mothers who were encouraging and who pushed them. It's the same with my daughters. If instead of two girls I had two boys, or a boy and a girl, things would not have been any different.'

I ask Nandini a question rarely asked of a male scientist—how much of a role did family support play during the high points of her career?

'I think the credit belongs entirely to the family, because at the end of the day they are the people you go back home to. They give you comfort and satisfaction. We are a very close-knit family. We enjoy being together, and have fun with the children. When they were very young, I did have some trying times while juggling tasks. I would run home at five pm like Cinderella at midnight. On days when they fell ill or things went wrong with them, life was difficult. Did I feel guilty for not being with my family all the time? Not really, but I did feel helpless at times. But like everything, it was just a passing phase. Nothing is permanent—least of all, our troubles. I am an eternal optimist. If there are problems, solutions would be around too. You just have to look for them. Ultimately, it's all in the mind. To feel stressed or happy is really one's choice,' she says with conviction. 'The positive side is that having a working mom made my children more mature, strong and independent. Now that they've grown up a little, they're not so dependent on me, and I can revel in my work. Take, for example, my coming back home yesterday after a twelve-day NASA trip. The family managed very well. My husband cooked a little in the morning. In the evening I have a cook who comes in. You do get a lot of help here, that's one advantage of living in India.'

She remembers being teased by her bosses about her 'support system' in the early days. 'My daughter was born in 1996. So after that, every time I had a launch and I had to go to the control centre, it would mean odd hours. The way the geometry and orbits worked out meant often working beyond daylight hours. So at those times I would request my mother-in-law, and she would come and spend ten or fifteen days with us, taking care of the kids. My bosses would say, "She's activated her support system, just like we call for extra ground support."'

MOM: Operations in Outer Space

Living right next to the URSC helped as well. 'When my daughter was born, we moved into a house from where I could walk to office. At the time, my parents and in-laws were both not in Bengaluru. But my brother was studying medicine here and was a great help. Later on, my parents moved here after my father retired. It's been easier since. Every time I have to go out and when my husband is not there, I call my mother or send the kids to her. It is basically the system and the family that supports you.'

Nandini gives a more recent example: 'I had been working hard for a big review and presentation. Sahithi [her daughter] was down with fever. I thought it was viral, but the next morning I had to rush her to the clinic since she was in a lot of pain. She was diagnosed with dengue and by the time the lab tests and consultancy were complete, I realized I was late. People at work rescheduled my presentation to after lunch. By that time, my mother reached the hospital and took charge. Such small incidents happen all the time. Men can also face such situations. In reality, women probably feel more pressure because running the household is still their responsibility. If that mindset changes it will be a big deal.'

Nandini was exposed to mathematics and science from childhood, learning about the constellations at a very young age. Her mother teaches mathematics to classes eleven and twelve, while her father, an engineer with a 'great liking for physics' is a consultant with the World Bank. 'Somehow, science was always there in my life,' she says thoughtfully. 'I didn't have that artistic bent.' Nandini completed her graduation and post-graduation in physics from Delhi University, cleared the Cambridge Society for the Application of Research (CSAR) national entrance exam and got a research scholarship with a Bengaluru laboratory. She also applied for a job at ISRO at the same time.

'In those days, since things were not yet computerized at ISRO, it took almost a year from the time I applied till when they sent me the recruitment letter. So actually, there *was* the option to complete my research. But I wanted to join ISRO,' she laughs, 'and after that I haven't looked back. I've been busy, happy and satisfied.'

What about parental pressure for marriage?

'They probably did worry like other parents,' she ruminates, 'but I got married early, at the age of twenty-two. My husband working at ISRO was kind of convenient. It was a very nice arrangement for both of us. I don't know what my parents would have said if I had insisted on focusing on my career first and marrying later. Everything just worked out in my favour. In any case, I come from a very non-patriarchal background. My father did many household chores. In fact, I preferred going to my father to do my hair for school.'

ISRO centres have a considerable number of husband-wife teams working together. Though there isn't any official data, this adds to the harmonious 'balance' all the women scientists claim to maintain between their work and home. A 'spouse in science' is an enabling factor for women with careers in science.

At the 'Sexism in Science' lecture, Nandini had said, 'Women need to do twice as well as men to be considered half as good as them.'

Did she mean they needed to work twice as hard at balancing home and office, or work twice as hard in the office?

'I wasn't talking about myself. I meant the general mindset that most women face and what I hear about in other organizations. Not only government organizations, but private ones as well. At ISRO, no one really cares if you're a man or a woman. As long as you deliver, they are happy.'

'In your 21-year, 14-mission career at ISRO, did you ever feel any kind of covert bigotry or bias?' I ask.

'Many times you could have individuals who harbour subconscious feelings like these,' she nods, 'but in terms of my career, promotion, getting opportunities—the way we were designated for very important missions like the MOM, or the significant projects I am doing today—I have faced no discrimination.'

So the work speaks for itself?

'It does.' She pauses and clarifies, 'There is no glass ceiling here, but you definitely need more numbers.'

~

Why are the numbers so skewed, with only a select few women becoming programme directors or deputy directors? In the absence of any statistics on women in senior positions, the general response of people in ISRO is that women are not treated differently from men. There is no separate categorization for 'women scientists' and hence no need for statistics. K. Radhakrishnan, Chairman, ISRO, however, acknowledged the crucial role of women at the core of the Mars mission, while also noting that they comprise 20 per cent of the total workforce.

Nandini responds to this analytically: 'Relatively fewer women joined as scientists in the 1980s, compared to the wider recruitment pool today. Logically, therefore, in ten or fifteen years' time, the percentages would be higher.' She also has another theory about why women are not found in the highest positions in the organization. 'It is very necessary for every woman to nurture the ambition of expecting to reach a top position. I have seen quite a few colleagues take voluntary retirement after reaching a certain level, after twenty years or so. This could be for personal reasons. I'm quite sure men do this also, but since the women are fewer, it gets noticed. If they had stayed on, they would have reached the top positions like Dr Vellarmathi—Deputy Director, URSC—and others. Now that there are more women in the middle management levels, you could probably see more in the higher management in the coming years.'

She elaborates that women do not necessarily need role models, but they do need examples to demonstrate that it is possible to reach the top, and that one shouldn't give up after the hard work of the initial years. 'You struggle in the eleventh and twelfth standard, come out with flying colours, get into a good college, continue struggling, complete a PhD and eventually get married. I've seen a lot of friends do extremely well and then give it all up for marriage, or to follow their husbands' careers. There is nothing wrong with that or with preferring to remain homemakers—if they are happy with that. But somewhere it is causing this problem of lower numbers. And then of course, there are many women who don't have the privilege or exposure. It's time they all knew the opportunities that exist, so that they can do whatever they want.'

~

Nandini's analysis is supported by the sobering ground reality for Indian women in STEM today. The perception that 'science is not for girls' is widespread in homes, schools, coaching classes for competitive exams, scientific/technical institutes and most importantly, within the public mindset. Despite this handicap, girls often outnumber boys in the science marks lists in the tenth and twelfth standard exams. In pure sciences at the undergraduate and postgraduate levels, girls are almost level with boys, comprising 40 and 35 per cent of the total strength, respectively. In engineering, the number of female students is lower: a mere 20 and 15 per cent of the total strength at the undergraduate and postgraduate levels, respectively.[30]

The latest statistics published by the HRD ministry in January 2018 headlined an impressive reduction in the gender gap at the university level in the last five years. The survey showed women's enrolment at the MSc level in mathematics, physics and zoology to be over 60 per cent, with chemistry registering a slightly lower 56.3 per cent. MBBS courses registered 99 women for every 100 male students. For BTech, the number was lower at 39 women for 100 men, with the IITs ranking much lower. Girls made up approximately 8 per cent of the total student enrolment at the twenty-three IITs across the country in 2017.[31]

This auditing falls by the wayside when we look at the data for girls actually going on to practice science, as opposed to those studying and teaching science. There is no updated information available on either the overall number of women scientists in various scientific organizations in India or their seniority. The few studies on women in science conducted by the Indian Academy of Sciences and the taskforce of the Department of Science and Technology (DST) are outdated—some of them are five years old.

Professor H.S. Savithri, Chairperson, Standing Committee for Promotion of Women in Science, tells me the reason for the lack of data. 'There is no single portal where information on women in science can be added and updated, so it is difficult to access information on women holding top positions as directors, heads of centres or faculty heads. Such websites, even if accessible, are not updated due to lack of knowledge, interest or time. Anyone searching for collated data would need to do their own research by going to individual institutions.'

The data that *is* available paints a dismal picture. As per the DST reports for 2017–18, 39,388 women or 13.9 per cent of the total workforce was directly engaged in R&D activities in April 2015. This is less than the Asian average of 18.9 per cent and significantly lesser than the world average of 28.8 per cent in 2014.

Women engineers at NASA comprised a meagre 20 per cent of the total workforce while 15 per cent were part of planetary science missions.[32] Back home, approximately 12 per cent women are faculty members at premier scientific institutions, while an even lower

10 per cent occupy senior roles—deans, heads of departments.[33] The same study estimated 30 per cent women working as project or division heads in ISRO, Defence Research and Development Organisation (DRDO) and the Department of Science, but no one at the top of the pyramid. No woman has ever become a director of TIFR, IISc, CSIR, any of the IITs or ISRO. Although, in July 2018, Dr J. Manjula took over as the first female director general at the DRDO.

This is more commonly known as the 'leaky pipeline' phenomenon, where the base of the pyramid never becomes wide enough to support the peak—indicating the continuous loss of women in STEM once they get to the top.

Apart from their accomplishments often being overlooked, women scientists are habitually excluded from informal scientific networks, where their male counterparts make contacts and gain visibility. They are also less likely to head advisory committees or attend scientific conferences as compared to their male counterparts.[34]

According to a 2010 report by the Indian Academy of Sciences (IAS) and National Institute of Advanced Studies (NIAS), 14 per cent women in science research had never married, as against only 2.5 per cent men; while 46.8 per cent women worked forty to sixty hours a week as opposed to 66.5 per cent men working fewer hours.[35]

When the objective of 'evaluating and enhancing women's participation in scientific and technological research' is transposed from a report into tangible, innovative policies plugging the leaky pipeline syndrome, women in science will truly arrive. Sensitively designed curriculum at the school level, with equal representation of women in the roles of scientists and engineers, is another way of breaking stereotypes.

~

At the URSC where Nandini works, there are 1,654 men and 549 women, out of the total technical force of 2,203, as of 2018. When the numbers are skewed, when science continues to be treated as a male bastion, who are the new icons for young girls today?

What advice would Nandini—as a freshly-minted role model—give to a school or college student in Meerut or Madurai, who has a distinct aptitude for science and wants a future in it? How hard is the road ahead for her and how best can she cope with the obstacles?

'If girls have the aptitude and interest in science, which is of course the prime requirement, they will need to work hard towards a single, persuasive goal. You won't always be successful, things won't always happen your way. But then that's the way to learn, isn't it? Don't get bogged down. There will be a few upsets on the way. We face them in our careers all the time. It's not that everything I do is always a shining success. Sometimes there will be disappointments. Ultimately, working hard is key, nobody achieves anything without working really hard,' says Nandini.

Speaking about the pure sciences, she declares, 'We really need to encourage this in our country. Very few people take up pure sciences. They do engineering, get a quick job and make easy money...'

'And after that they do an MBA, so their engineering degree is wasted and then they take up jobs in corporate banks,' I add lightly.

'Or they sell soaps,' smiles Nandini. 'That's one way of looking at it—they're happy that they are making money.'

If university teachers were paid as well as MBAs or people selling soaps, if teaching was transformed into a lucrative job that people competed for, then the best minds would teach children and encourage their interest in research, she feels. 'It should be made very difficult to secure the post of a lecturer or even a high school teacher, to ensure you get the most brilliant people to ignite the students' curiosity. A good teacher makes all the difference.'

At Delhi University where Nandini studied, the top students would do physics and the next layer would do mathematics and chemistry. But in Bengaluru, everyone goes for the engineering seats. If they don't get in, only then do they do science.

'I met a few Indians in NASA, all from IITs. They go there and score well in theory and zero in practicals, because our exposure to labs is almost nil. We need to upgrade our labs, get the best minds to teach the children and expose them to different institutions like ISRO, or have people from here go and teach,' says Nandini. 'I've been giving lectures in schools and colleges regularly, even before the Mars mission. I don't like to say no, even if I'm busy, because I know that children will be interested.'

Nandini's workplace facilitates visits from school and

college students on a regular basis. Sometimes up to 200 students visit the space exhibition in the main building in a single day. National Science Days are celebrated at the URSC every 28 February, with quizzes and competitions designed to motivate students to pursue careers in space.

'We need more workshops for children across all states, so they actually see what science is—we should make small rockets and show them. Build a scientific temperament early on.' Nandini shares the example of how NASA's Johnson Space Centre provides a hands-on experience, where one can go into a simulator and feel the weightlessness of being in space. 'They also have an entire group of people looking at outreach programmes, and a dedicated NASA channel. We are a cash-starved country, so we can't do the same, but science needs to be promoted,' she says emphatically.

'True, otherwise today's urban children watch *The Big Bang Theory* or the *Star Trek* series—that's the extent of popular exposure to science,' I offer.

'*Star Trek* was one of my favourite television serials,' Nandini laughs. 'I've seen all the *Star Trek* films too. Watching the serial—and Captain Kirk—on Doordarshan was a family ritual.'

'What is a typical work day for you?' I ask.

'Days are pretty hectic—early morning chores, sending the children off to school and college, driving into office by eight-thirty am, working till six or six-thirty pm and then catching up with the family back home. I don't spend too much time teaching my children or following up on their studies. They don't like that. I plan the next day, sleep early and get up early,' she says. 'During mission times, of course, it is a fourteen- or sixteen-hour work day, especially if you're the mission director like I was for Resourcesat-2A [a remote sensing satellite launched in December 2016].'

Does she have any other interests or hobbies?

'I used to play the violin. I learnt classical music, but the last five or six years have been crazily hectic. I don't have much time to attend concerts. Somehow, with two children and this crucial time in their career paths, we have cut down on going out and social gatherings. When they were younger, it was more fun: with no exams to worry about, we'd just pick them up and go. Now this twelfth standard period, especially in our country is very crucial. There's always some exam or test going on. The kids' schedules don't match, and sometimes my husband's and mine don't either,' says Nandini ruefully.

~

At Nandini's unpretentious duplex in Jeevan Bima Nagar, I ask the two attractive young ladies sitting in front of me, 'Who is your go-to person in times of trouble?'

Twenty-one-year-old Sahithi and 16-year-old Mahima reply in chorus, 'Mom'. Sahithi adds that while both her parents are strong influences, her mother has played a larger role in helping her choose her career: medicine.

'A lot of people think that having a working mother is really difficult and they feel something is missing. But I don't think either of us experienced that, because we lived close to where our mom works and we'd go to day care in the afternoon hours till we reached the eighth standard. And when we were little, unless she had

Those Magnificent Women and Their Flying Machines

a launch coming up, she would be home by six pm, so we would spend most of the evening together. We had weekend fun too. I think we spent a lot of time with her. If anything ever happened to us, she was always a stone's throw away and would be there in five minutes. So we never felt we missed out on anything. It also pushed us to be independent. We now feel we can do so many things by ourselves.' Neither of the girls asked Nandini for help with homework, for instance. Sahithi manages on her own or relies on her grandmother while Mahima prefers checking in with her older sister.

Sahithi remembers accompanying her mother to work occasionally, in the early days when the rules weren't so strict. When Sahithi was around four and a half years old in 2001, Nandini recalls, 'There was a very important operation in the night for which I had to go. There was no place to leave her—my husband was out of town and I had no family in Bengaluru at the time—so I decided to take her along. I planned it out, giving her dinner, carrying a blanket and pillow, so I could make her sleep on one of the chairs in the office. Enroute, we had a mother-daughter chat. She asked me, "What will you actually do today?" I told her there is a satellite that goes up to one of the stars that you see, I will tell the satellite to turn on its instruments and it will do its job and collect the data. She was not quite convinced and asked, "Is your office in open air, will you go to the terrace to tell the satellite, won't you need to see the sky?" I told my colleagues later and we all had a good laugh.'

Did she sleep peacefully after that?

'No, I wanted her to sleep but there was so much activity around. We have this protocol that we follow during launch—whoever sits on the mission director's chair and takes charge of the operations becomes MD [Mission Director], and others address people according to designation: MDComputer, MDNetwork, MDShadnagar [ISRO data reception station, National Remote Sensing Centre near Hyderabad]; that's how we normally talk. Sahithi found that very amusing, so for the next few days she went around saying "MDComputer" while calling us. She watched the whole launch and eventually slept after one am in the resting room.'

In recent years, stringent security measures have barred family members from launch events, including the Mars mission. There were other ways, however, in which MOM crept into the Harinath household during its two-year duration.

Nandini recalls one such incident. 'It was the time when we had just finished the earth-bound manoeuvres for MOM around midnight, and the engine hadn't fired for the TMI burn. (The TMI is a major operation where the spacecraft is pushed into the Mars transfer trajectory with the velocity required to leave the earth's sphere of influence). The chairman called a meeting at three-thirty am to discuss what to do next.

'We had been in office all day and all night in stressful conditions, and by three am, we were all really drained. But since the chairman had the energy to do a review, no one even thought of sleep. Later that morning, I had to attend a PTA meeting for Mahima, who was then in class seven. So, I had called her earlier and told her to request her teacher if I could come to school at two pm. Now Mahima does not talk much about what I do at ISRO. She merely said, "My mother is doing something for the Mars mission and working late, so she will come later." The astonished teacher immediately told her she was doing well and there was no need to disturb her mother!'

How much of an advantage is it for both parents to be in ISRO?

'That's the best part,' says Sahithi enthusiastically, 'my dad is in the testing aspect of the satellite, so they do all the work before the satellite is launched and my mom is in the commanding part, where she is busy after the satellite is launched. So they are busy at different times. It's very rare for both to be working on different missions and busy at the same time, so we are happy.'

'You know, my husband is a typical man who may not really volunteer to do everything at home, but when it's required he steps up,' says Nandini. 'He has his own interests—he's out on a trek now in the mountains—but if I say I am going on tour, he will take complete charge. He will even cook.'

The easy camaraderie Nandini shares with her daughters is obvious. So is the daughters' quiet pride in their unique mother, who is a role model in her own home.

'I don't really know how I ended up doing my MBBS,' says Sahithi, 'when from the third to the twelfth standard, all I ever wanted to be was an astronaut! That was my dream all along. You know those essays you have to write as a child on what you want to be when you grow up? I always wrote "astronaut". I wanted to be like my mother.'

'So you'd be able to crack that diagram I keep encountering all the time and struggle to decipher—the one with the Mars Orbiter and concentric circles depicting orbits?' I challenge her.

She laughs, 'I think I would, hopefully, since I was always interested in everything my mother does, especially during the Mars mission. I'd ask a lot of questions and she would explain how many things could go wrong and that they were trying to make sure they did not.'

At the time of the Mars mission, Sahithi was studying for her twelfth board exams and the competitive entrance tests. 'It was a pretty hectic time for my mom, and I was studying hard too. So we would set an alarm and both of us would get up at four am. She would do her work and I would do mine. So even though she wasn't actually teaching me anything, just by being there for me, she made it so much easier to concentrate on my studies. In between I could take a break, talk to her... we'd have an early morning coffee. That was really nice,' Sahithi recalls.

Did she feel that the Mars mission competed with her board exams for Nandini's attention?

'Not at all,' she declares, 'I feel that because she was so busy with it and had so much work to do, and I had so much work to do, we did our work together. If she hadn't got up with me at four o'clock every day, I might not have done so either.'

Sahithi shares an incident that happened when she

was in medical college. 'One day one of my friends came up to me and showed me a Facebook post about three women who work at ISRO, and their achievements. I told him the one in the middle is my mother. After that, the entire college found out my mom worked on the Mars mission. So if any posts came up, they would promptly send me a link.'

'So you hadn't boasted about your mom at all? Not just the Mars part but that she works in space research?'

'I think we did. My close friends knew,' she admits with a grin.

Mahima chose to write a speech on MOM for an assignment, taking nuggets of information from her mother that her classmates couldn't collect from Wikipedia.

I ask the girls how they relax with their mother. Do they watch films, or go to concerts?

Both girls giggle. 'My mom is not a movie person at all. I am really into sci-fi movies, but she has this killer line she uses on everyone: "We don't need movies for all the drama and suspense. We see drama every day as part and parcel of our work." She did see *The Martian* though, since everyone from her workplace was going for the special preview. While we were fascinated by *Interstellar* and *Gravity*, her reaction was "hmm".'

Nandini interjects, 'I do watch films once in a while, but TV puts me off. I watch the news to catch the headlines.'

'Mom's more of a stay-at-home person, chatting with us on a Sunday, ordering lunch from outside, reading the newspapers and magazines Dad subscribes to or shopping online,' Sahithi says fondly as Mahima nods in agreement.

It is now past nine pm and conscious of delaying their dinner, I move on to final questions. 'Do you see your mom as an example of gender equality?'

Sahithi responds with alacrity, 'Absolutely. When my friends and many of my juniors in school and college read about my mom and tell me how inspirational she is for them, I realize how lucky I am that I got to live through all of that. We don't have to look too far for role models, there are people who are with us right now and she is one of them. My mom is such a strong person, even emotionally. One of her strongest traits, which I understood gradually over the years, is the way she handles situations. She never loses her temper and keeps her cool if anything goes wrong. I hope to be like that some day.'

Nandini, and other women scientists at ISRO, are such relatable role models because they are 'normal' people juggling an impressive number of balls in the air on a daily basis. They are not reclusive, nerdy scientists unconcerned with the outside world.

'A lot of people think taking up science in India is really hard, that it's a tough life, but it's not impossible. People like my mom did it. So there should be many people who can also do it. If they want to contribute, there are so many ways to do it,' Sahithi says.

I am curious about Nandini's own role models, the women who inspired her.

'I've wracked my brains, but I don't think I've come across any woman of Marie Curie's stature, for example.

You have so many men, C.V. Raman, Homi Bhabha, Vikram Sarabhai, Bose...so many others. But I am struggling to find names of women,' she says regretfully.

'Maybe one day soon, one of the MOM women will be a role model for young girls,' I tell her.

She laughs, adding, 'A while ago I attended a function in Gujarat for Nobel Laureates and there was just one woman among eight or nine men who had won the Nobel. So it's a fact we need to accept.'

'Your face lights up when MOM is mentioned. Was your husband ever jealous?' I quiz her.

'He was in the testing department for MOM. Since it had to be completed in fifteen months, all resources were pooled and everyone contributed in some way or the other.'

Asking a space scientist about what lies ahead is exciting, since the answers could well be beyond our imagination.

'I am part of a project team that combines work for the Mars follow up, for Venus, and for an asteroid mission. A small committee comprising a handful of people have made a proposal and sent it to headquarters. We hope it comes through.'

'What will you do on Venus?'

'Venus is a lot more difficult than Mars since it is a little hostile, especially the temperature which is really hot. It has an atmosphere which is difficult to plunge through, a very thick cloud that you need to pierce. You cannot use optical sensors. You need microwave payloads, which means more weight, more power and more money. It is a different technology. Today the government policy is to back projects that have immediate social applications, where the public can see the difference made to the common man and the way he lives. The emphasis is on satellite data for navigation and disaster management. But of course, space exploration is always there, be it Venus or any other planet.'

If there's any truth in the Roman poet Ovid's words, 'Venus favours the bold,' Nandini's next space odyssey is on the right track.

~

Car journeys in Bengaluru can seem endless and exhausting. Sometimes, they provide a glimpse into a person's innermost thoughts as conversations, like the traffic, take unexpected turns. I am travelling in an Uber to Insight Academy in south Bengaluru, where Nandini Harinath has been invited to give a lecture to senior school students. This is part of an ISRO tradition, suggested by former URSC director T.K. Alex to all senior scientists, urging them to go back to their hometowns and schools and talk to the children about what they do as space scientists. It is Bengaluru-born Nandini's turn today.

The slender, spectacled scientist is impeccably dressed in a blue printed silk sari with earrings to match, a small bindi and a mangalsutra around her neck. As we pass a temple enroute, she remembers the evening before D-Day, the Mars MOI on 24 September 2014.

'Since Prime Minister Narendra Modi was scheduled to attend, they had sanitized the entire Mission Operations Complex at ISTRAC and cordoned it off, with dogs in place and everything locked down. We were told we could not leave the campus. Whoever was inside had to stay there till the next morning for security reasons. No one was allowed in either.

'We had finished our work. There was no uplink to be done as everything had been done a week in advance. So, after dinner and the last review by the chairman [K. Radhakrishnan] who cleared it, all we had to do was wait and watch. Around nine-thirty pm, our boss and mission director, Kesava Raju, Ritu and I decided to go visit a nearby temple.

'We were wearing new badges that indicated we had been checked in. I asked the security person at the gate if we could go out for a few minutes. The three of us went to the Shiva temple nearby. It is a small, roadside temple that I used to stop at and pray before all important events, since it is on the way to ISTRAC. At nine-thirty pm, obviously the shutters were down. But it has this grill through which you can see the deity even from outside. So we all prayed and then quietly returned by ten pm, feeling happy and relieved.'

'You mean, God will take care...?' I ask.

'Yes,' she responds. 'On all important days I have a few such rituals—I wear clothes that are given by my husband or my parents. I visit this temple on the way. I was once travelling with Kesava Raju and MOM Project Director S. Arunan on one such important day, and I was wondering how to tell them that I wanted to halt at the temple. As we came near the junction I asked them if I could stop for five minutes to pray. They said, "We will also come with you." So all of us got down and prayed.'

'The normal image of scientists is that they aren't religious...'

'Well, it is one thing to believe in God and that there is somebody who is above all of us. It is another to be superstitious. We don't believe in superstitions.'

'Who could understand the mysteries of the Universe, however inexplicable, and the likely existence of a higher power, better than space scientists...' I agree.

Nandini laughs, 'I always feel there is some practical reason behind everything we do, and many times just that five minutes of prayer gives you a lot of peace of mind. It also helps to keep the mind blank and not think about anything for a while.

'A calm temperament and nerves of steel are absolute prerequisites for working in space,' she stresses, 'especially for those who are in operations. Space is very demanding. You cannot afford to make a mistake. You can do it in the rehearsals. That's why we have so many simulations and testing. We go through precise and minute checking, and then there are reviews, so if anything is missing, it will be pointed out.'

Since this *is* rocket science, I am curious to know what happens if anyone makes a mistake or buckles under pressure.

'Of course, mistakes happen, since we are all human. Sometimes you just blank out and sit and wonder what happened. You can't imagine what could have gone wrong, when you think you have done everything. But then you learn from it and move on. And there is always a team, people to back you up and to support,' Nandini smiles.

'No one can really do what they want totally on their own. They are accountable for what they deliver to the higher authorities, thanks to a rigorous review mechanism.

'Reviews take place regularly, and if there are anomalies, we have an immediate review within a few days where you need to submit a report and set out in detail what happened. Otherwise also, we have performance reviews of satellites—the different milestones at different times during the project that you need to cross. Each needs to go through a review to get to the next level, and those are also time-bound,' she explains.

Such rigour is common to all projects in ISRO's sixteen centres across India, and reviews are conducted by different people at different times. It could be your immediate boss, a director or the ISRO chairman, depending on the project. Peer reviews are conducted by members of the board or committees formed for each project. These are people from different disciplines and specializations such as thermal, power, control and others. Each scientist simultaneously works on and is responsible for multiple projects at various levels.

'I had a review recently where the mission director was my boss's boss! I was three levels down and designated as just a young manager when the satellite was launched, fifteen years ago. Now that everyone higher than me has retired, I am supposed to take care of the satellite [Resourcesat-1, a remote sensing earth observation mission launched in 2003] as long as it is alive and doing well,' Nandini tells me.

'You also have to deal with time clashes—sometimes you have an important review coming up and something else comes up that is managed by different people who obviously don't know what is happening in your other projects. They only care that you finish theirs! And this is happening more and more these days,' she adds. MOM was an exception, however. 'For those eighteen months the priority was only Mars. MOM was a very big team effort, with maybe ninety designated persons and at least five people working under each of them.'

The good part of her job is that it gives her a lot of freedom to work and execute her ideas. 'Nobody micromanages or forces you to do things in a particular way. They have the confidence that the work will be done and do not need hourly or daily follow-ups. You are only answerable during reviews,' explains Nandini.

What happens when a review isn't quite favourable?

'Many times reviews are tough and people don't agree with your opinions. Either you take action and convince them you'll get back with further analysis, or you just accept whatever they say,' says Nandini frankly. 'If someone has more experience, one goes by that. Also multiple solutions can exist for a problem.'

The fact that ISRO projects are team-based and someone else is always available to step in and take over, helps women scientists during their maternity leave. However, the gaps do affect their progress up the promotion ladder. Nandini is thoughtful before confiding, 'You know, we say maternity leave doesn't affect promotion, but in some cases it certainly does. Although there are many dimensions to the promotion issue. Some women on maternity leave have got their promotions on time, while others have not. Some men have worked hard and yet, they have not got promoted.

'There are a lot of women who are very senior, Anuradha T.K., N. Vellarmathi, Seetha Somasundaram… but so far no one is heading a centre.'

'Well, I hope you go ahead to become the chairperson of ISRO one day,' I tell her.

Nandini's eyes twinkle, 'I don't really have such aspirations. What matters is merit, and one should have been a very senior programme director or director of a centre to stand a good chance. Otherwise it will be like a reservation. If women have to demand that respect from society, they have to stop asking for concessions.'

At this most interesting point in the conversation, the car halts in front of our destination and we are engulfed by the unmistakable sights and sounds of a lively school day. Before we can ask one of the children for directions to the lecture hall, the principal of Insight Academy, Mary D'Souza, and science teacher Kajal Roy greet us. We are escorted to the assembly hall where 109 students are already seated and waiting, amidst a familiar buzz of whispered conversations.

The hall's side entrance overlooks the playground, and younger students pass by occasionally, shooting curious glances at us. The projector is set up and the presentation is cued. Nandini begins with a few basic questions to determine the level of knowledge and science background of the ninth and tenth standard students. Isaac Newton kicks things off with the three laws of motion. All of the students know about Newton and the apple-falling-from-the-tree story. Nandini explains the law of gravity in the context of space, satellites and the work she does at ISRO.

She explains what an orbit is, along with rockets and the types of satellites launched by ISRO. Communication, navigation, weather forecasting, earth observation, remote sensing, military and scientific satellites—Nandini traces their different orbits and applications and explains why people need them today.

From television or radio broadcast signals to the forthcoming indigenous navigation system (NavIC) replacing the American GPS coordinates, from world weather predictions to tsunami/flood/earthquake alerts, from keeping watch on our borders to studying the solar system, she explains that satellites consciously or unconsciously impact our daily lives a lot more than we know.

As everyone listens to Nandini, I am tempted to ask questions myself to clarify long-forgotten basics of physics. The students seem to understand it all fairly well. She explains how the placement of the satellite in various orbits depends on the functions it performs. An earth observation satellite is kept in an orbit of 250–1,000 km above the surface of the earth, communication satellites range from 1,000–20,000 km and so on. Travelling to a different planet in the solar system—for the scientific satellites—means escaping the earth's gravity in the hyperbolic path.

Nandini tackles the parts of a satellite next—an antenna for communication and tracking, power through batteries charged with solar panels, thrusters to keep moving the satellite every now and then, sensors that play the role of eyes to pinpoint the satellite's orientation with respect to a star (stars are fixed in space) and finally the processor, the control system that is the brain of the satellite.

'What are thrusters used for, ma'am?' asks a young boy in the front row.

'Thrusters are used for manoeuvering a satellite. If you are in an orbit and need to change the speed or the orbit, you need to fire the thrusters. They are like the small rockets you fire on Diwali. You fire the thruster and you get a force in the opposite direction—again, Newton's laws.'

'You can stop me and ask questions at any point, because if you keep your questions till the end you might forget,' she adds.

Most of the students pay rapt attention; many take notes in their notebooks. A few gaze vacantly, perhaps dreaming of outer space or the forthcoming lunch break. Nandini shows animated videos explaining how geosynchronous satellites work and why they are essential. 'You all know that there is nothing stationary in space. Everything in space is moving—earth, planets, satellites. So our satellites have to be synchronized with the earth's motion... This is the only way we can use the satellite constantly for our communication needs. Therefore, as the earth makes one rotation in twenty-four hours, the satellite should also rotate at the same speed. To do this, it has to be placed in a particular orbit in a particular way.'

Nandini tells the students how satellite pictures of

devastated areas during the Uttarakhand floods in June 2013 helped rescue teams to identify the areas with maximum damage. 'Remote sensing satellites, on the other hand, give faster, repetitive observations of objects without actual physical contact, by using gravitational, magnetic or electromagnetic force fields. Scientists can estimate, for example, if an agricultural crop is healthy or not through infrared radiation... Google maps are all made of satellite images—each and every nook and corner of the globe can be mapped with the satellite.'

Since the Mars Orbiter Mission is the most anticipated part of the lecture, Nandini keeps it as the finale. 'Why would you want to go to Mars?' Nandini asks the students.

One of them immediately replies, 'To find if life exists.'

'Absolutely. And you would want to actually go and explore it rather than looking through a telescope, right? Our population is growing and our resources are dwindling. So one fine day—a few decades from now—there could be a reason to go and probably start living on other planets. Then there is the million-rupee question: "Is there life on Mars?" This is why ISRO decided to launch the Mangalyaan mission,' declares the deputy director, operations, MOM.

'The satellite has to go to Mars which is also moving in an orbit along the sun. It has to leave at a precise time and velocity and in a particular direction, so that it can meet Mars at the exact time that Mars is supposed to be there, within a precision of two seconds,' explains Nandini.

'Then the satellite has to be inserted into the Mars orbit, get captured around Mars' gravity, which is the most important burn because if that does not work, the satellite would become a flyby. It would go past Mars and get lost in the solar system. It was a do-or-die moment, you couldn't get it back,' Nandini says to the students listening in awe. 'Targeting something 650 million km away, within an error band of 50 km, was likened by NASA jet propulsion laboratory director, Dr Charles Elachi, to hitting a golf ball from Bengaluru into a bucket kept in Los Angeles, while the bucket is moving.'

The students laugh and applaud as the 45-minute lecture draws to an end. As the assembly disperses after a heartfelt vote of thanks, several students, a large contingent of girls among them, gather around Nandini, asking her rapid-fire questions.

'If there is an interference in the signals of two satellites, if they intersect, what will happen?'

'In that case you simply won't get the signal.'

'Will there be any variations that will give wrong information, ma'am?'

'Whenever you get a signal from a satellite, you have to validate it. There are some minimum parameters to validate if the signal is coming from your satellite or not. If the validation check fails, the software rejects it. Same thing happens in a satellite...each satellite has a receiver and a decoder. The decoder decodes the information it receives. So if the information does not match what it is supposed to receive, then it is rejected and you won't get any signal. There is no chance of spurious information,' explains Nandini.

MOM: Operations in Outer Space

'Ma'am, to pursue a career in space science, what is the career path one should take?' a girl from class ten asks.

'Space science is very vast...it can accept a person from any field. So you have material sciences, you have physicists, mathematicians, engineers. You need all types of people to make a satellite. You need to do a good graduate or postgraduate course in science or engineering from any good institute and you can then apply for a job at ISRO. How many of you want to do that?'

Almost all of the 20–25 starry-eyed students—boys and girls—raise their hands. Nandini beams and instructs them to study hard. After patiently autographing several notebooks, she leaves for the principal's office, where there is just enough time for tea and a quick photograph, before we head back to the URSC.

In the car, I ask about her most exciting missions and her face lights up. 'In 2007, I was the operations director for a mission called SRE-1 (Space Capsule Recovery Experiment), where the satellite was actually designed to go and orbit for ten or twelve days and then fall into the Bay of Bengal, from where we had to retrieve it. The objective was to demonstrate that we had the capability to come back. This was extremely important also for the human space programme, for the reusable launch vehicle.'

According to the official ISRO website, the 550 kg space capsule—SRE-1—was designed to demonstrate the technology of an orbiting platform for performing experiments in microgravity conditions.[36] SRE-1 travelled 'round the earth in a circular polar orbit at an altitude of 637 km.'[37] After completion of the experiments the capsule was de-orbited and recovered.

'It was our first reentry mission and as thrilling as the moon mission, Chandrayaan-1. Yet it was accomplished without much publicity, so most people didn't know about it then, or even now,' says Nandini. 'We launched on 10 January 2007 from Sriharikota, using the PSLV C7 rocket and reentered on 22 January, and it was awesome. We needed a totally different technology to get back into the atmosphere. Because of the friction when entering with a very big velocity, you would be burning at more than 2000 °C. Therefore, you need a good thermal protection system to keep you safe. Then there was the guidance and navigation part to bring the space capsule down, and a flotation system to keep it afloat after impact. We had to inject the dye on impact to identify where it had fallen and then the balloon would come out to keep it afloat. There was real drama and suspense in this mission.

'We had to collaborate with the coast guard and the navy because ships and helicopters with their own tracking systems were placed where the space capsule was expected to fall on reentry. We had to re-orient and fire the engine to break and change path, to decelerate and get into the earth's atmosphere. That was a very critical burn. It went off beautifully, with textbook precision,' says Nandini with pride.

'...When it entered the earth's atmosphere, it was in a blackout zone for about two minutes. The minute it came out of blackout, we got the signals confirming that we have successfully reentered. That was a dream moment.

'Then it went down just below the horizon of Sriharikota Range where the SHAR station could not see it, where the helicopter systems were supposed to track it. Suddenly, there was no signal and we were all left wondering what happened. The helicopters and ships began searching and as the minutes went by, we were all praying. But there was just no sight of the capsule. The sea is so vast, it isn't possible to visually search all of it.

'Someone then had the brilliant idea of taking the last GPS position, before the loss of signal, and writing a quick software to extrapolate that curve till it touched the sea. They did the calculations, we conveyed it to SHAR who conveyed it to the helicopter and it was found right there—majestically floating within a few metres' accuracy,' exults Nandini.

'The flotation system had opened but the commercial beacon that was supposed to send the signal ran out of battery and failed. Everything else went as planned. People cried and hugged each other when we got the information—the first time that an object from space had come back to earth,' she continues.

It is anyone's guess what ordinary citizens in the area felt on seeing an object hurtling down from space, but Nandini assures me that the coast guard would have explained the situation adequately.

'Your life really *does* have enough sci-fi moments that you don't need to see science-fiction films.'

She laughs. 'Someone told us yesterday that there will be a Bollywood movie on women scientists of ISRO. Did you hear about that?'

Chapter 3

MOM: The Payload Performers

Since, in the long run, every planetary civilization will be endangered by impacts from space, every surviving civilization is obliged to become spacefaring—not because of exploratory or romantic zeal, but for the most practical reason imaginable: staying alive.

—Carl Sagan

On a hot, summer morning, I find myself in the air-conditioned waiting room at the Space Application Centre, ISRO in Ahmedabad. Following ISRO protocol, my mission to meet MOM's 'payload women' needs to be routed through the Centre Director, Tapan Misra.[38]

As the morning stretches on, I begin to get restive. After almost ninety minutes I am herded into a sizeable, tastefully decorated office. A group of men are on their way out. The director is visibly busy—fielding calls and signing documents at an enviable pace—but he welcomes me warmly, signalling a break to the secretary hovering nearby.

Tall and burly, dressed in a colourful, bright green shirt matching the indoor plants, Tapan Misra is an unusual leader. An alumnus of Ramakrishna Mission and Jadavpur University, he has edited poetry magazines, curated film clubs and is a regular blogger. He speaks plainly, openly sharing his views in his impeccable Bengali-accented Hindi.

He also provides names, introductions and access to nineteen other women scientists at his centre, instead of just the MOM 'payload ladies', whom I originally intended to interview. 'Meet others also,' he exhorts. 'There are a lot of women doing excellent work. I always say if you want a job done quickly give it to a man, but if you want a job done very well give it to a woman.'

Elaborating his point, he says, 'We tend to overlook women. This exists in management everywhere. One can call it dominance or animosity of men towards women. But I feel that men are susceptible to "prime vision". If they decide they do not want to see something, they will not see it, even if it is in front of their eyes. Take, for example, this pen that lies here. If I decide the pen is not there, even if I search I still will not find it. This is the male brain that doesn't want to see a woman perform. But, of course, they are there. Without them, the organization does not run.'

'As the director of SAC, Ahmedabad, what have you done to change that perception?' I ask.

'Ultimately, it is a game of numbers. I made a very subtle change by making it mandatory to incorporate 20–30 per cent women in every committee, from junior to senior levels. Earlier, people would think of including women as representatives only on sexual harassment committees. Then we started identifying women, designating them for presentations, lectures, professional interactions, and putting them in charge of organizing professional programmes.'

Misra also claims to have erased the demarcation

between the 'manly' and 'womanly' tasks of men and women at SAC. 'When I took over as director, women's roles revolved mostly around paperwork. I started putting them in satellite integration. I was ruthless: if this meant women had to stay up all night, so be it. And today, in this campus, you will find women at the forefront of many integration jobs that have tough delivery schedules,' he observes with satisfaction.

I ask about the gender disparity in ISRO: 'Why is the number of women scientists and technical staff less than 20 per cent, with so few women managing to reach the seniormost positions?'

According to Misra, 'Women who are joining today have a better chance to go up. Earlier they had to face much more opposition. Today the concepts have changed. I will not be surprised if we see a woman chairperson in another ten years.' He adds, 'Opportunities exist for [women] to go far—one may agree that this far is not far enough if you take a look at the women in high posts at ISRO today. They joined thirty years ago and are probably paying the price for a non-level playing field at that time.'

How level is the field today?

'I would rather see if the playing field is hostile or not. We have made it much less hostile, but ultimately the playing has to be done by the player. There is only one type of goal in football, whether it is men or women shooting it. But you cannot give the male footballer a 1 kg boot and a female a 5 kg boot. That difference we have removed,' assures Misra.

In the absence of hard data, it is difficult to verify the

progress of women scientists in the organization over the five decades of its existence.

~

There is no record of ISRO's first woman scientist, though serving scientists like Tapan Misra acknowledge the pathbreaking work of women such as Saroj Prabha, inventor of the algorithm for the IRS 1-A (India's first remote sensing satellite) or Deepti Rastogi, whose talk-back experiment laid the foundations of the Edusat satellite for distance learning. Ramadevi T.S. is another classic example of how high a woman scientist can climb at ISRO. The Kerala-born engineer started work as a technical assistant in the Vikram Sarabhai Space Centre (VSSC) in Trivandrum in 1970, and retired as its deputy director in 2010.

Ramadevi says, 'Though I put in four decades at ISRO, I still feel I might have become more than a deputy director. But I don't regret my slow career progression, mainly because I had a settled life in Trivandrum, where I could bring up my children and my husband could concentrate on his career. Achievement of a position, not promotion alone, is rather difficult in a highly male-dominated workplace. For me, it was my technical competence combined with the ability to get along with people, which worked. I have been assertive without being aggressive. This helped me get through my career… ISRO has changed a lot in terms of creating more opportunities for women, as well as enhanced awareness of their rights.'

When she was starting out as a 22-year-old engineer, she was not allowed to go to Ladakh with her male teammates to test a gadget they had developed together. This is less likely to happen today. She urges young girls to take up science and technology, not only to help them in their chosen careers but as an excellent training ground for life.

~

Misra admits that women have to work slightly harder than men for recognition. They have to perform better than a man to get the same promotions. Though several ISRO women scientists hold a differing point of view; they actively dislike the 'woman' tag since the quality and volume of work is no different from their male colleagues. 'Women have a wider view. They bring another dimension to the work, their approach is distinctly different in the management of people and their work. As long as this difference is there, we should celebrate it,' maintains Misra.

The work done at ISRO is fast becoming iconic in the way it is followed and appreciated by ordinary citizens of India, owing to the favourable press and social media attention. 'I'd say our reputation is becoming a problem, since now people join ISRO with starry eyes,' says Misra in mock despair. 'Recently, I told some new recruits that their arrival is similar to what people experience the day after the marriage, when the house is dirty and one is faced with all the cleaning up. Suddenly, the feeling of achievement is over but all the small, day-to-day jobs have to be done.' They soon realize that working at ISRO, the premier science research organization in the country, means discipline and meticulousness.

Moumita Dutta

Project Manager, Methane Sensor, MOM

Inside IRS (Indian Remote Sensing) IC Clean Room, I try and get comfortable in my 'bunny suit'. The blue overall doesn't quite fit, but I manage to fasten the velcro panels, wear the matching blue cap that completely covers my hair, and sit down to put on protective paper booties. Sliding one foot at a time over the bench, I finally emerge on the 'clean side'. Putting on the paper mask and standing under a vent for sterilization with pressurized air is the last requirement before entering the 'clean room'. It is here that one gets a firsthand glimpse into the wondrous world of the payload scientist. Today, the Project Manager, Methane Sensor for MOM, Moumita Dutta, is showing me how and where she worked for the eighteen-month duration of the Mars mission.

Gowned and capped like me, Moumita leads me through the brightly lit, high-ceilinged room with no windows and no link to the world outside. Had the room been silent, it would have been akin to a sensory deprivation chamber, but the hustle and bustle of morning activities was reassuring. A 'clean room' filters out particles lesser than 0.3 microns to ensure that the room and the instruments inside the laboratory are contamination-free. This is one way to take care of the lifespan of a payload and decrease its chances of failure.[39] For Moumita, the room is probably her second home. She has spent up to eighteen hours without a break inside its tall walls several times during the Mars mission.

She proudly shows me the model replicas of the three payloads she worked on—the methane sensor for Mars, designed to measure methane in the Martian atmosphere and map its sources; the Mars colour camera, which gives information about the surface features and composition of Mars and also monitors dynamic events and weather; and the thermal infrared imaging spectrometer, which measures thermal emission and maps the surface composition and mineralogy of Mars.

Designed in India, this equipment had exacting prerequisites—it had to be compact, lightweight and cost-mass-power-effective, since the mission mandate was to use minimum fuel, and it had to be ready in fifteen months.

After taking a closer look at the MOM payload replicas, my attention veers towards the rest of the room—the various work stations in front of complex machines and incomprehensible scientific equipment, and the scientists and technicians in blue overalls busily going about their tasks. Nobody gives us even a cursory look as we make our way out.

Once we have left the clean room, and our protective suits and shoe covers behind, Moumita shares the details of her MOM quest as we sip tea. 'It was very satisfying, an experience of a lifetime. Be it the technical, administrative or managerial parts, or even how to work under tremendous pressure, the Mars mission has taught me a lot.'

'How did you get selected for it?'

'I had worked on planetary and earth-observing

missions like Chandrayaan-1 in October 2008 and Oceansat-2, a climate and environment satellite launched in 2009. So I had the necessary experience for this kind of payload. The payload we built for Mars was a totally new, first-of-its-kind sensor,' she tells me.

Slim and elfin with large, expressive kohl-lined eyes and a serious countenance, Moumita reminisces, 'Back then I used to go home to sleep for a few hours or to have dinner. The rest of the time, I would be in the lab for sixteen to eighteen hours continuously. We had to test the payload in different temperatures commensurate with the space environment. Other teams would also be waiting to take their measurements. This meant that we would often be in the lab for two or three days with no thought of food or sleep. We would just concentrate on the measurements. Our focus was to make and deliver a very good instrument and, since it was new, we had to figure things out. Every day, I would only be thinking of the next day's work while going home. I would see those optical components in my dreams, and worry about how I'd aligned them. If I had a problem, I'd spend the night thinking of how I would solve it the next day so that the work didn't get held up. We had no time for pre-planning or making mistakes, but often, it's at such times that your brain is very active.'

'Weren't any scientists at senior levels able to help out?' I ask.

'Seniors are definitely available but when a problem arises you cannot depend on anybody else to give you a solution. It's your responsibility,' she says.

'What was the toughest part of your involvement with MOM?'

Those Magnificent Women and Their Flying Machines

'Everything was a new learning for me. We used many new optical components, including a special optical filter that was the key to giving us information about methane. All components were commercially bought. The most suitable ones were sourced and had to be brought here [SAC, Ahmedabad]. We had practically nothing in hand. We looked very carefully for each and every component, factoring in the time it would take to reach here and most importantly, if it would give the desired performance. We had eight or ten components of which two or three were totally new. No one knew how to test them'.

When there was barely a month left to complete the project, Moumita found out that the instrument she needed to test the camera was not available. The entire test plan—devised by her—was based on that instrument. 'I was in a tight spot, but fortunately, I had a back-up plan which was even better than the original. However, even with the back-up, I didn't know if the testing laser would arrive in time. Finally, after fifteen days, it arrived and I made a new set-up quickly and we were able to test the entire payload successfully.'

Workplace challenges during the MOM project meant not being able to visit her ailing mother in Kolkata, even for a couple of days. 'I felt guilty at that point, but my mother herself stopped me, saying that I was working on a project of national importance. For eighteen months, I hardly attended any family function or thought about my family, focusing wholly on completing my assigned tasks. My husband went abroad for some time and I barely got any time to talk to him. But all this didn't upset me at all. MOM isn't the only project where we face such situations. In every project there is always a phase—lasting perhaps a couple of months—when I work at this pace. I don't want to glorify it as my sacrifice. When you are working for the nation, you should be grateful that you have been chosen to contribute. What can be more satisfying than that?'

I ask Moumita if she dreamt of being a space scientist as a little girl.

'I used to read a lot of science fiction as a child. I wanted to communicate with aliens on another planet,' she admits.

While pursuing an MTech from Calcutta University, Moumita read about ISRO's upcoming Chandrayaan-1 mission to the moon in *Anandabazar Patrika*. This turned out to be a defining moment for the applied physics student. 'I kept imagining how the spacecraft would be going to the moon, how lucky the scientists were—all from our own country—who had the opportunity to work on this prestigious mission. After completing the course, I joined Calcutta University as a research fellow for a brief period, before getting an interview call from ISRO. At the same time, I received an offer from Dublin to do my PhD in fibre optics technology—something related to my MTech project. I didn't think twice before rejecting the foreign offer and joining ISRO. When I came here in January 2006, my dream came true when I discovered I was going to work for Chandrayaan-1 on two payloads—the Hyperspectral Imager camera and the Terrain Mapping camera,' recalls Moumita, her eyes shining.

Like the other women scientists interviewed in this book, Moumita too makes it a point to acknowledge her family's support. 'My parents, my in-laws, my husband never put any restrictions on me. They never told me, you can't do something. They never complain when I stay long hours in office for my projects. I have completed ten projects here. Every time a project deadline approaches, I have to work very late... My family is doing the best they are able to do. Now it is up to me and my capabilities to see how far I go.'

If the words beneath the obviously genuine, heartfelt sentiments tell their own story—of concessions and permission from family members—it will be a while before people, both men and women, accept hardworking women in the workplace.

Minal Sampath

Project Manager, System Integration, MOM

At the Sensors Development Area in SAC Ahmedabad, Minal Sampath declares, 'I have always wanted to be the first woman director of a space centre.' But she doesn't intend to take any shortcuts. She wants to toil hard, learn and jump through all the hoops to get to the top. 'Only then will I be happy,' she says.

Minal's ambition was evident from a young age. In the eleventh standard, when she watched a satellite launch live on TV with wide-eyed admiration—the space scientists in their white garments, the thrilling moments of the countdown till T0, and the final take-off, she felt, 'This is the place I should be in. This is the work I want to do.'

MOM: The Payload Performers

Many years later, Minal stood in the exact same position during the Mars Orbiter Mission launch, marvelling at how her wish had come true. The 39-year-old scientist is emphatic that if you want to achieve something with a sincere intent, there is no doubt that it will be realized one day. 'Now I know why our school teachers would make us write essays on "If I was the Prime Minister"... They wanted to enhance our imagination, give us a vision. At that time, we used to think of it as a five-mark question to be answered...

'Children, especially college kids, come up to me today and ask for guidance—"How do we go about doing something good for the country? We are bouncing around like free atoms... Someone suggests we do an MS, someone an MD." I tell them to start dreaming, to analyze themselves, since no third person can do that for them. Others can only point you to a path, *you* have to walk it.'

She is equally firm with anyone who shies away from the inevitable hard work required for a career in science: 'If you feel you'll have to study much more in IIT or engineering—yes, studying is essential. Only high marks can lead to tremendous opportunities. You have to make a choice. If you think you can get away with lower marks somehow, then you'll have to run extra hard later on to grab the same opportunities. So you can run now or run later, but the running will need to be done.

'Make small goals but underline what you want to do with your life. It is not as if dream kar liya toh so jao [Now I've had the dream, might as well sleep]. You still need to work hard to make it come true. Science also

Those Magnificent Women and Their Flying Machines

teaches a person to analyze life on a day-to-day basis in significant ways,' says the space scientist, who is a gold medalist in electronics and communication.

Minal has an air of lively self-confidence about her. Obligingly, she recounts stories about her childhood and the lessons she learned growing up in the small town of Rajkot.

In the second standard in school, she had failed her Hindi exam. 'I spent the whole day crying with my face in my hands. My mother kept saying, "Don't worry, I know you studied. It's not the end of the world, you will have more chances." I assured her from that day on I would take my studies very seriously, and I did.'

No one in Minal's middle-class Gujarati family had studied in an English-medium school. So when she faced difficulties in class five, her mother, a mathematics school teacher, considered shifting her back to a Gujarati-medium school. 'I told my parents that in time I would do fine. That was my second lesson—never give up. If I had switched, I would have always felt that it is okay to give up easily,' says Minal matter-of-factly.

Her first brush with gender discrimination within the education system came in class twelve. 'Our mathematics teacher used to divide the classes into girls' and boys' batches, because he believed boys were better than girls at maths. That was my first trigger and I asked, why this differentiation? Why assume that girls don't know mathematics?' Minal says indignantly. But she took it in her stride and moved on towards an engineering degree at Nirma University in Ahmedabad.

'Engineering was mechanical in nature and I liked

all that, but at that point I did not know which field to choose. I promised my father that I would excel wherever he put me and he chose electronics. Throughout the four years I did very well, got excellent grades and also met my future husband who was my classmate,' she says with a laugh.

In the seventh semester of her final year, they both filled in the application forms for ISRO on the last day, scrambling to collect college transcripts and other documents. The interview call in Bengaluru meant spending Rs 9,000 for a one-way flight ticket. Minal's father insisted on her flying and even accompanied her, so she could be 'fresh and rested', instead of undertaking the tiring train journey. Minal remembers how nervous she felt almost eighteen years ago, as if it was yesterday. 'I told my father the airfare would be wasted if I didn't get in, and he told me to just do my best and leave the rest to God and destiny.'

She describes what happened next with her trademark self-assurance: 'There was a panel of ten or eleven people who interviewed me. At that time there was no written test like there is today. It was based on an interview and college grades of the shortlisted candidates. They asked questions related to the basic fundamentals we had studied and gave us some problems to solve—all related to communication. My first boss was among the panelists. He later told me he had never seen a freshly-minted graduate answer so confidently before.'

Minal and her husband, Rohit Sivakumar, began work at ISTRAC in 1999. Her early assignments were using communication satellites to provide access to medical aid and education for people living in remote areas. One by one, the opportunities started flowing her way through her 'wonderful journey' at ISRO, which continues to this day.

'I always say ISRO has given me a lot and it is my duty to give back. I should do things to help people. I see people chasing personal goals—mera job lag gaya, bachcha ho gaya, badi car aagayi, bada ghar aagaya, abhi settle ho gaye, life mein saving kar lo [I've got a job, a child, a big car, a big house, I'm settled, I just need to save now]. But is this the ultimate goal? You have to do something good, some research for the common man, for the country,' declares Minal.

Contributing to the Mars mission was the perfect way for her to fulfill both goals—serving ISRO and the nation. Mangalyaan proved to be the most rewarding and challenging mission of Minal's space career, pushing her beyond limits she could not have foreseen.

She describes the pressure she and her colleagues were under from 2012–2013: 'We had two big projects happening at the same time, with launches scheduled for 2013 within a few months of each other—INSAT 3D[40] and MOM.

'One team for INSAT-3D had already left for Kourou, French Guiana. My performance review was also scheduled during that time, but I didn't worry about it too much. I thought the (MOM) payload has to be sent off on priority and surely the review bosses will realize the work I've done on that. So I tackled all three together along with frequent travelling. If you look at my itinerary for the last six months of MOM, I have travelled from Ahmedabad to Bengaluru at least ten–fifteen times, because all the testing is done at URSC.'

As a member of the INSAT-3D project—which shared team members with MOM—Minal would send off emails from Bengaluru to her boss at Kourou late at night. 'I would ask my boss how things were progressing and he'd admonish me for being in office past midnight. I'd tell him the testing was still going on. He would give me further directions, all well past midnight. It was a hectic and stressful time,' she recalls. 'I would face physical and mental exhaustion often, but what kept me going was the feeling that I have the power to overcome, that it is just a matter of time, of allowing it to pass.'

As soon as testing for MOM at URSC got over, Minal would rush home, taking the first available flight at any hour of the night. 'My son was very young at the time, just three or four years old, so I was needed on the home front as well. At one time, when he was ill, he wanted his mother beside him. But I consider my payloads to be my children too, and I was needed there as well. Fortunately, since we live in a joint family, my husband and in-laws came forward to help. My mother-in-law, in particular, is very active and helped out a lot. She still does, every day,' she says appreciatively.

'I help her with the kitchen chores, get my son ready for school in the morning, drop him off and then run to office. I am just in time or a few minutes late sometimes, which nobody minds, fortunately. Then I plan out the day's agenda, work until six pm or later, if there's a project running. Go home and take my child skating, get dinner organized and sleep by ten-fifteen pm. In between, I call home in the afternoon to check if my son has completed his homework, since his grandfather is very soft on him. And then I am told he doesn't listen to anyone except me. Why should he?' Minal says with a laugh.

'What about the dark and difficult periods, did you ever feel you can't cope?' I ask.

'Often, despite your best efforts you don't get results. But then I listen to my inner voice which tells me to hold on, that everything will be fine. My vocation and commitment to my job gets me out of tough situations. I am able to do this as I know I have put in the best possible effort and needn't worry about the consequences. The same thinking applies to problems at home as well. I've found that there are many similarities to problems on both fronts.'

Does she have any regrets about not doing enough for her family while working for ISRO?

'Once I take a decision, there is no regret because I have assessed it and made the best possible choice at that time. Even if I go wrong, I learn from it. If you do something with good intentions then there is a supreme power that takes care. The nation is always top priority and so is family. One needs to prioritize and work with meticulous back-up plans,' Minal avers.

The office peon brings tea and I continue the interview. 'What exactly is the work you do—now and during the Mars mission?' I ask.

'I am a system integration engineer. Payload development is a multidisciplinary activity involving many expert teams working in electronics, optics, thermal, mechanical and system integration. Each team has a defined role, and they have to send their deliverables to the system integration team, which is the last lap of payload development. This is a phase where the risk factor is very high. As part of the system integration team, my role is to integrate various payload subsystems[41] and make them work together. My journey begins from integration of the payload subsystems at SAC Ahmedabad to URSC Bengaluru for payload integration with the spacecraft, to the launch pad at SDSC-SHAR for integrated spacecraft level final testing before launch and then finally, post-launch testing once the spacecraft is in the intended orbit and stabilized in the space environment. This is the end-to-end journey of any system integration engineer.'

Minal explains that under normal circumstances, payload development is like a 'relay race': activities are sequential and follow a pre-defined path. First, an engineering or qualification model is made, if there is a new design. After it qualifies, a flight model is made. In the case of MOM, due to the time crunch, the two models ran simultaneously: while the engineering model was going through all the rugged extra testing, flight model development also began, and all the major testing was done during the integration phase.

'So the minute you say "on", you have to be very sure that everything is as per what you have planned and discussed orally.' Before the crucial moment of powering on, Minal would always pray: 'Bhagwan ka naam leti hoon, Sri Ganeshaya Namah karke [In the name of Lord Ganesh, I begin],' she confides with a laugh.

Completing all the tests was Minal's task in MOM as the project manager of system integration. It was a formidable one, with no quarter given. 'I always say space takes care of you if you perform correctly. But if you make just one blunder, leave out one observation, it will finish you—it is the worst demon. So it is all meticulously planned and a lot of thought goes into it—"If this wire fails, what happens? If this component fails, what is to be done?" We do a *lot* of testing.'

The time and resource constraints also meant delivering within a specified, inflexible deadline and long, tiring days inside the closed-off, windowless clean room. 'There was no excuse for not completing a job, because then you would miss the deadline. If you didn't deliver you were out.'

The continuous time spent in the clean room—not seeing the sun for days on end, not knowing or caring if roads near your home are flooded with rainwater—also takes its toll. 'Even if there's an earthquake or floods, you won't know what has happened, because it [the clean room] is seismic-proof.' She jokes that MOM's hectic schedule turned her hair white (she has a full head of jet black hair). 'The testing phase of any payload development goes on and on, 24x7. There is no way you can break from work, go home and come back. You have to check the data yourself.'

Minal considers the payloads her 'babies'. 'I really feel there is life in them when I touch them, when I'm doing the wiring. Maybe I am more emotional or psychic, but that's how I am. How can you leave your child and go away?'

MOM: The Payload Performers

I ask her if perhaps space scientists who work on such landmark missions deserve more recognition than cricketers or film stars.

'The general public doesn't know us. For them a scientist is still like an alien. Before the Mars mission, we had Chandrayaan-1, which got some media coverage. But MOM went directly to social media and a lot more people got to know our work. In fact, we asked our bosses, "Why have you put all this online? If something goes wrong, the public will blame us." But they felt that people should know what we are doing.'

Thanks to the MOM spotlight, it's not just the public but even her own eight-year-old son, Siddharth, who can now visualize a different-looking scientist. Not an Einstein or an astronaut in a spacesuit, but someone like his mother—a normal, regular person.

'I tell him what a real scientist does—she forgets to eat or sleep, she doesn't know if it's day or night. So when Mama works on such projects she won't come home on time, she won't eat properly, she may forget she's a Mama.'

Minal has also inculcated gender sensitization lessons for her young son. 'Recently my son and I went to buy a birthday gift for his friend, a girl. The shopkeeper asked who the gift was for. When I told him it was for a girl, he showed us craft items, drawing books and Barbie dolls. My son also told me to buy a soft toy. I asked him to gift her a Meccano set and see how she would make a better helicopter than him.

'I have seen many girls who are shy and scared. It's as if they think somebody is going to eat them up. Maybe their environment has been like that from the beginning. They have been told to keep quiet. Even in my own house, when my voice was loud they would say, "Dheere dheere". Though, my parents never differentiated between my brother and I in terms of schooling. They gave us both total independence.'

'On the job, are women scientists treated any differently from or by their male colleagues?' I ask.

'If you want to take up challenges, people are ready to give you opportunities. Whether you are able to cope is up to you. Ask for your rights, fight for them, don't feel shy. You have to take the first step. I have my own dreams and I don't fear anyone. You should be true to yourself.'

What's the next big thing for Minal now?

'I am part of Chandrayaan-2 with two payloads. We went ahead with the Mars mission, based on our confidence from Chandrayaan-1. Now MOM will help in realizing Chandrayaan-2. Every mission enables us to discover and develop new technologies. We still have a long way to go. We need young minds to come in. This is happening because they are happy to work for the country, but they need to be guided, especially girls.'

Chapter 4

The Vanguard Veterans

In village schools, there was no equipment [for scientific experiments]. So I wrote down a credo, which was that the objective of the science programme would be to help children realize that science is everywhere. Science is in the kitchen, in the village pond, science is in the bicycle, science is in the flora and fauna....

—Professor Yashpal, scientist and former SAC director, ISRO

The open-air canteen at URSC, Bengaluru is unusually noisy at one pm on a sunny August day, as throngs of hungry schoolchildren pick up dosas and lunch thalis from the harried servers at the counters.

The tenth class boys and girls of Seshadripuram High School, Yellahanka, are visiting the space exhibition on display at the ground floor at URSC. From displaying satellite systems, scaled models of satellites, including replicas of Aryabhata, INSAT-2, Chandrayaan-1 and Mars Orbiter Mission, to information about satellite technologies and the history of ISRO—the exhibition is a comprehensive overview of India's march through decades of space research.

Though the exhibition is open to all, students from various schools and colleges form the bulk of the visitors. ISRO encourages such visits to motivate young minds to take up a career in space research. The young minds I talk to—a smattering of uniform-clad girls willing to delay their lunch—are suitably impressed and motivated. Shilpi announces her intention of pursuing a career in space; she is awestruck by the Clean Room where they observed satellite integration in progress. Several others claim they will study hard to get good marks to qualify for a science career. The class had to write a report about what they learned at ISRO, so they have all spent more than an hour taking notes.

'Do you know the names of any of the women scientists of the Mars Orbiter Mission?' I quiz them. The girls do not, but all of them assure me solemnly that they will follow their dreams even if they are not sure what they are just yet.

Seetha Somasundaram

Programme Director, Space Science Programme, URSC, Bengaluru

'What would you say to a young girl studying science in a small town, who has a clear interest and aptitude in the subject but is discouraged by parents or social environment from making it her career?' I ask 57-year-old Seetha Somasundaram, one of ISRO's seniormost women scientists.

'It is at the school level that you realize if you're interested in science or not, because that is where you get the basic training, where you study all subjects. So I would tell schoolchildren—boys and girls—sometimes people around you can find out what you are interested in. Your friends, siblings or teachers may tell you, "Hey, you're good at this." Parents are usually more concerned with your upbringing rather than interests. But nowadays, there are many parents who identify their children's interests. For those children who do not have any outside support, I would say just observe yourself. Wherever you find that you are doing things without studying, that is probably where your aptitude lies. Ask yourself if there are subjects where you do not struggle, where you forget yourself. Therein lies your ability.

'When you're working on a problem, even a simple problem such as solving a question not in the books or the curriculum, and you solve it and get pleasure out of it, that is the first indication that you have an aptitude for research and science,' Somasundaram declares. 'Research is all about the problems out there and how you choose whatever is appropriate to solve it… If a girl is keen and good at science she should be encouraged to pursue it. It is not as difficult a career as people make it out to be.'

'How easy was it for you?' I ask her.

Seetha describes the academic environment prevalent in her time (more than three decades ago), how she countered patriarchy and her exultation at achieving economic independence.

She speaks unhurriedly, but with emphasis. 'My father was an engineer working with All India Radio and therefore, not against science or engineering. But he was still not very forthcoming when it came to me pursuing an engineering career outside my hometown. He didn't object directly or say I should get married and not study further, but he couldn't quite digest my living in a hostel. Since I had done my final schooling in Delhi and I was interested in science, I joined BSc honors at Hindu College. But I insisted on doing my MSc from IIT Madras, and since my father considered IIT hostels to be safer than other hostels, he allowed me to try campus life. When you are a day scholar your space is limited, but while living on campus we were free to work till night-time in our labs. They would give us a key and sometimes, we would go back to work after dinner. I loved my teachers who gave me that freedom,' says Seetha.

'I finished my MSc in 1979. I had applied for ISRO while I was still studying. I was recruited in the following year, joining the science group called Technical Physics. Today it is called the Space Astronomy group. In those days, space was an upcoming field. ISRO was a new organization and I thought I could do well here.' Seetha felt that a PhD could wait till her work demanded it. She wanted a job in research and development. 'I always tell students, especially girls, if you want to have your say, be economically independent. For example, you cannot say you don't want to pay dowry for your marriage unless you are independent.'

Somasundaram's chronicling of her initial years as a young woman scientist at ISRO reflect the organization's early stumbling blocks and hard-won triumphs. The early 1980s were a time when Indian satellites and launch vehicles were being developed, when certain constraints (financial and otherwise) meant reshaping scientific goals and experiments, adjusting time-frames and persevering despite setbacks. Somasundaram joined the Space Astronomy group, initially working on developing rocket payloads before moving on to write project reports for a science experiment on a satellite.

This experiment was organized in collaboration with scientists at the Tata Institute for Fundamental Research (TIFR), who had been working in the field of X-ray astronomy. They planned to build an X-ray astronomy science satellite with a 60 kg payload. However, the limited carrying capacity of the early Indian launch vehicles meant that this project was put on hold.

Instead, the Astronomy Group was given the opportunity to work with a 5 kg payload for the intermediate Gamma Ray Burst[42] experiment in the mid-1980s. 'At that time, GRBs were some of the newest celestial phenomena observed. Nobody knew where they were coming from. Even today, so many decades later, it isn't exactly pinpointed but we are able to identify some of the sources causing Gamma Ray Bursts,' recalls Seetha.

This 5-kg experiment was flown in the Stretched Rohini Satellite Series (SROSS). The first two SROSS satellites, SROSS-1 (March 1987) and SROSS-2 (July 1988), failed to orbit due to failure of the launch vehicles. The third one, SROSS-C (May 1992), launched by the ASLV-D3 successfully got into orbit (the first satellite to have done so) but it was short-lived.

The next in the series, SROSS-C2 on the ASLV-D4 (May 1994), hit pay dirt and GRBs were detected. 'One of the foreign satellites detected it and we asked them for the light curve to compare and then, exactly on cue with the time difference, it was visible. We could finally detect the Gamma Ray Bursts. We were so happy as a team,' reminisces Seetha.

The ten years that it took for the ASLV-GRB experiments to achieve victory (the Government of India had sanctioned Rs 19 crores for the project in July 1982)[43] were crucial. Seetha looks back on them with a mixture of nostalgia and regret. 'Probably, if SROSS 1 and 2 had gone into orbit, we would have got the GRB earlier, and then, the space science programme would have grown faster. Since we had these initial setbacks, we began an Optical Astronomy programme in the interim, using the ground base observatory at the Indian Institute of Astrophysics (IIA). We used to go there for optical observations. At that time, this meant that you had to be at the telescope floor all night, monitoring in the dark. Nowadays, it is all remote controlled and you can sit in a lit room and see it... But in those days there were very few women astronomers—parents would hesitate before sending you. In our team of fifteen, there were two or three women, and I was the only one doing optical observations at that time.

'All my colleagues and the people at IIA were very cooperative. The work atmosphere was very good and we all had a nice time. Perhaps this is why I feel that when you enjoy your journey, you aren't worried about whether your career is going up or not. Even if the progress is slightly slow. This often happens in science—you like your work so much, you're probably not saying, "Hey, that person got a promotion." Someone who is keen on a career graph might have shifted to another project, but I didn't. Having initiated something I wanted to see it to fruition,' she says emphatically. She started work on Astrosat soon after, but that too took a longer than anticipated to complete.

'A programme that a scientist is working on for some years may suddenly be sidelined, as another gains priority. It is possible that men shift faster, to follow the important programme. This could be the reason for less women at the top of the pyramid,' reasons Seetha, 'but the social restrictions on women are also undeniable.

'Once you complete your BSc, BTech or even MSc in your early to mid-twenties, your parents start thinking about marriage. Even if they "allow" you to work, they specify a time limit of three or four years till the inevitability of marriage. So the job is actually considered a stopgap or a stab at brief economic freedom. Those who *do* try and balance work and marriage, and continue with their jobs have to take a break if the spouse moves away.'

Many of Seetha's colleagues took up bank jobs so they could get transfers. The forties are peak years for work promotions for men, while at the same age women are on the mommy-track, managing children and career deadlines, and possibly losing out.

'You need to join early enough and stay on long enough,' says Seetha.

~

The Department of Science and Technology (DST) has a comprehensive scheme known as KIRAN (Knowledge Involvement in Research Advancement through

Nurturing), which encompasses all science programmes to help women advance their scientific careers and empower them 'to break the break' of maternity leave and the child-rearing years. They facilitate a smooth reentry into the mainstream. The money provided under these Women Scientists schemes is generous: Rs 20,000–55,000 per month—and the age limit is from 27–57 years.[44] Off the record, however, senior women scientists are skeptical about the arbitrary implementation and successful outcomes of initiatives like this.

Survival for women scientists is linked to a list of demands generic to all working women everywhere—safe spaces and a professional atmosphere at work; daycare at the workplace; transportation facilities; flexible timings; one-third representation on important committees, including grievance cells for sexual harassment; the support of parents and in-laws; a spouse who also has a career in science, who will understand and empathize; a mentor system; training courses for women to get into middle and senior management; inclusion of gender sensitization in the education system and gender auditing.

Gender discrimination continues to be pervasive today and was especially dominant thirty years ago during Seetha's days as a young scientist. 'In the 1980s, when I joined ISRO, there were very few women in the engineering and science groups, so seniors questioned whether women would be able to handle the rigour involved in this work. They believed that women would work for limited hours because they needed to go home.

They would not stay back if there was a problem. They would not be committed. Gradually, the women who joined ISRO dispelled that feeling. Of course, some women left but that happens everywhere, even among male scientists.'

Seetha claims that if she had switched to the remote sensing or communication programmes, she might have moved ahead much faster. Both fields witnessed a lot of activity at the time, following Vikram Sarabhai's dictum to use space research for social applications. Science missions are only now reaching that stage with Chandrayaan 1 and 2 and MOM. 'In science missions you always have to find a niche and do something that has not been done earlier. You cannot do it [make a new discovery] every two years or so. It takes time,' she states matter-of-factly.

Seetha outlines the unique challenges in doing space science experiments: one cannot correct or fine-tune them mid-way, unlike ground-based experiments, and one has to rigorously dispel apprehensions about their efficacy. Most importantly, one has to start them off as piggyback experiments—an experiment on a satellite not meant to be a science satellite, where you share time with other payloads—often taking a backseat.

In the IRS-P3 satellite launched in March 1996 by the PSLV-D3, astronomy targets were observed in the three-month monsoon period when remote sensing could not be done, since the remote sensing payload was the primary payload. After this astronomy piggyback experiment was successfully accomplished, the space science teams were asked to work towards a space science satellite. This was a dedicated mission confined within the capabilities of the ISRO launch vehicle and satellite and yet, configured experiments that were scientifically contemporary. They came up with Astrosat, a single satellite to study and make simultaneous multi-wavelength observations of celestial targets like stars and other astronomical objects.

Astrosat, launched in September 2015, is the first satellite operated by ISRO as a space observatory, in order to facilitate a deeper understanding of the universe. Fine-tuning Astrosat continues to occupy Seetha, which she terms as one of her career's 'big thrills', along with the initial GRB sighting and the Mars mission. Astrosat-2 is in the study phase along with Aditya, the solar observatory slated for 2019–20 to observe the sun through optical, UV, X-ray wavelength payloads.

'Do you feel unfulfilled in any way?' I ask Seetha.

'Professionally, I would want regular science missions at ISRO. I think the time has come where there should be an opening for one science satellite every five years. They still will not be as many as remote sensing or communication satellites, but they are important. Research is not only about dealing with needs, it also deals with how you can aspire to do bigger things. If you aspire to grand science, only then will you develop many things on the ground that will be useful in other ways. Science is demanding, which puts people off initially. But it enables us to be more capable. There is an obligatory rigour to contemporary science—you have to work towards it and you need to invest time. People who want quick results may not be suited to good science,' she says reflectively.

'All achievers succeed because they stretch beyond what they are asked, without anyone telling them to. This is how individuals and teams grow. Women scientists are better recognized when they work in teams,' she adds, citing the fields of atomic energy and space.

On the domestic front, Seetha shares how her husband, a former ISRO scientist, was supportive of her work and how her in-laws looked after her infant son five months after she gave birth.

Seetha invites youngsters to join space research with the following scenario: 'Space research merges science and technology. In the future you may be able to do a lot of state-of-the-art things in space, better than on the ground. Space research is for the future generations. In fifty years from now there might be space travel. We might go to the next planets.'

N. Vellarmathi

Deputy Director, Payload Data Management and Space Astronomy Area, URSC, Bengaluru

Natarajan and Ramseetha will remember 27 April 2012 for a long time. As well-wishers, neighbours and friends felicitated them with shawls and flowers at their home in Tamil Nadu's Ariyalur district, the proud parents basked in the glory of their daughter N. Vellarmathi's success.

Just a day earlier in Sriharikota, Vellarmathi, the project director of India's first indigenous Radar Imaging Satellite (RISAT-1), witnessed the culmination of more than eight years of work as the 1,858 kg satellite soared into space—ticking off all markers at the expected times. She later addressed a packed press conference: 'All women are equally capable and have good potential which should be properly utilized.'

As local organizations planned grand functions in honour of Ariyalur's 'daughter' and students of Nirmalaya Girls Higher Secondary School—Vellarmathi's alma mater—waited to meet their role model in person, perhaps not many knew about the single-minded focus shown by the scientist thirty-six years earlier, when she had started out.

Her parents may remember her struggles though. It was her strong-minded mother who convinced her reluctant father to allow her to pursue an ME (Master of Engineering) degree from Anna University, Chennai. Their plan for their brilliant daughter had been straightforward—with a good bachelor's degree in engineering, she would get a job nearby, visit the parents on weekends, marry a 'good boy' locally, settle down and avail their help later to care of the children. Such a mindset towards a daughter's future continues to prevail within the average Indian family. But Vellarmathi was having none of it. She argued with her parents past midnight, overriding their fears of 'finding a boy for marriage equivalent to her' and pushed past the opposition from conservative relatives to forge her own path.

Today, she calls it nothing less than a 'breakthrough'. She is grateful to her parents for taking the risk in letting her pursue her dreams. 'The trend is now shifting with the new generation. Girls are doing well even if they still don't totally match up to the boys,' she says.

'Parents still have fears about girls being able to take care of themselves—"If you are attached to a man after marriage, he will take care of you"—but slowly this will change.'

Vellarmathi finished her post-graduation in electronics and communication, cleared ISRO's open panel interview for her first job and moved to URSC in Bengaluru. I cannot resist asking her about her marriage. She smiles. 'After about a year, my parents started looking for a boy for me. My father came to Bengaluru with three addresses and, unbeknownst to me, conducted on-the-spot interviews with the boys. He liked Vasudevan, an MSc Agriculture rank holder who worked in Vijaya Bank (Vasudevan has now retired as deputy general manager at the same bank). They [the groom and his family] came, "saw" me and we got married and settled in Bengaluru.'

The 57-year-old Vellarmathi belies her quiet demeanour. She is feisty and refreshingly candid in her observations, which often contain a hint of homespun wisdom. She tells me about the early years at ISRO. 'Initially ladies were not accepted immediately because our skills were not proved. We had bookish, theoretical knowledge from college, so a lot depended on which department we were posted in. For the first couple of years we did face gender disparity. Men generally accept and appreciate women if they have similar examples of working women in their own families. I don't know how many of my bosses' wives were working at that time. The trend shifted later...

'In those days, a satellite would be launched perhaps

once in a year, or two years, unlike the multiple launches today. Multiple reviews, including technical reviews, used to take place simultaneously in four–five rooms.' Vellarmathi would sit in a corner with other junior scientists and watch in awe the fluid ease and expertise with which people answered questions. 'On one occasion a project director came in from another review room, casually wrote out the answer everyone had been looking for and left.' Impressed, Vellarmathi resolved to emulate that knowledge and confidence as a leader.

Almost three decades later, as the project director of RISAT-1, Vellarmathi was asked, 'What would happen if it [the RISAT-1 launch] was delayed by a week? Would corrections be needed?' Her response was as impressive as the one she had admired all those years ago. 'If that happens, I will happily go home and relax with my family, since I am 100 per cent confident we have completed everything. Nothing has been left to chance. Just tell me whenever it is to be launched and I will come one day earlier and launch it.'

On 26 April 2012, everything went as per the meticulous planning and sky-high expectations. The complex, hardware-intensive RISAT-1 was launched successfully in the first attempt. Vellarmathi considers it one of the most rewarding assignments of her 33-year stint at ISRO, deeming it on par with any international mission. She describes the final moments, as she sat in front of the computer controls animatedly. 'T-plus-12 minutes: the rocket is going to take off. When it takes off, the satellite is no longer in your hands. Even if you want to correct anything, you can't. It has started going up. I imagine the satellite...how it will be travelling. My immediate concern is that the rocket has to perform well, so I look for first-stage separation, second-stage separation, and then the heat shield separates. I feel overjoyed: "Oh, my satellite is in space!" What is the sun exposure? What is the temperature? Which side will be facing the sun? My mind is full of these questions. After this bit is successful, I look for the next event. Solar panel deployment and then the following markers. Once the solar panel gets deployed, that means 90 per cent relief. Then my experts track the spacecraft, audio-video feeds come in, people speak over the phone and public address system. There is tension till that moment. And what a jubilant moment it is.

'When you teach your child a dance sequence and make her practice it several times, you will have the confidence to watch it on D-Day, because you know this moment is written. With the numerous rehearsals, the ground experience held good. Everything went off on the dot,' she says, the relief still evident in her voice as she describes the events of five years ago.

'You should listen to taxi drivers in SHAR if you want to know about spacecraft performance,' she tells me with a chuckle. 'At Sriharikota, several taxi drivers come to the operations centre daily, so they overhear engineers talking in the car after completing their shifts, discussing the areas that are causing problems.

'Everyone must have been discussing how everything looks fine with RISAT-1 and that they would have to come to the centre for just four or five days. So my taxi driver tells me, "What, madam! Everyone booked taxis for 15–20 days. Now people are saying everything will be over in four or five days!" So that was the certificate that things were fine. We packed up in five days and returned to our normal routine. There was such a heavy demand for slots in that particular satellite that it was fully booked three months in advance just like a marriage hall! It served the country with its direct applications.'

The ISRO website introduces the RISAT-1 as 'a state-of-the-art Microwave Remote Sensing satellite carrying a Synthetic Aperture Radar (SAR) Payload…which enables imaging of the [earth] surface features during both day and night under all weather conditions.' The SAR enables 'applications in agriculture, particularly paddy monitoring in kharif season and management of natural disasters such as floods and cyclones.'[45] Vellarmathi explains its functions: 'The concept is based on the eye principle. For example, when light falls and is reflected, you see the image. But in a dark room, you cannot see this. To see a picture you need light. The immediate question then is, how will you see the picture if there is no light? The RISAT-1 satellite provides the answer—it sends its own signal, irrespective of dark or light, cloud or rain. So you know what you have sent, you know what you are receiving. You have to check what the difference in that is, find out the properties, the particular terrain of the particular image and map it. Then you will get the image. So it is continuous twenty-four-hour imaging, irrespective of whether there is sunlight, clouds or rain. RISAT-1 involved a lot of manpower and money. It was a big, magnificent satellite.'

The Vanguard Veterans

Vellarmathi's pride shines like a beacon. 'We used to have on-orbit observation meetings (when the satellites are in space) and reviews where many satellites are discussed. For this satellite, seniors would write a "dash"—which meant no observation [i.e. no anomaly]. That was something to be proud of, a collective effort paying off. We had some twenty-five or thirty people giving their best, often working till four am with complete dedication. Meetings would be called at two am—we would all sit together, try and find solutions to problems ourselves. We would work weekends too: think over a problem on Saturday, then on Sunday, after a good lunch, when everyone at home was resting, I would come to office at two o'clock and work till eight pm. And then on Monday morning, I'd put up option one, two, three, four on the board. They [the bosses] would select one and we would implement it. Problem solved.'

'What about the rest of your team? Did you face any hurdles as a female boss—were your male subordinates respectful of your position as project director? Did you face any sexism?' I ask.

Vellarmathi is dismissive. 'In my 33-year service record so far, I have never faced a problem with men—either in my day-to-day activities or in my reviews. That is why we were able to grow to this level, and will continue to grow. When I took the post of project director of RISAT-1, I encountered and dealt with all kinds of people—good and bad, intelligent, confusing, argumentative and conflict-creating men as well as women. If you are capable and have potential, you will be accepted. There is no discrimination.

'My boss once told me, "You know, Vellarmathi, I never treated you as a lady here, never thought that I should not load you with too much work because you are a lady." I told him, "Sir, I too never behaved like a lady. You called me at three pm, told me to go to Ahmedabad at six pm and I ran from office to Ahmedabad. I never gave any excuses; I did whatever had to be done immediately." Even in one's personal life, one has to accept some conventions—if ten men are sitting and chatting in my house, I cannot go and sit with them. They may not feel comfortable. One has to consider all these things, but in my mind there is no gender difference.'

In earlier missions, such as the Technology Experiment Satellite (TES) launched by PSLV-C3 in 2001, Vellarmathi was responsible for overseeing all activities from spacecraft assembly to the successful launch and operation of the satellite. 'It was the first time a woman was given this type of responsibility,' she says. Dispelling the apprehensions that she would not be able to work nights, she made sure she was the last person to lock up and the first to receive the key the next morning. 'When you work this way, the others also become automatically focused. How did *I* manage? My husband is a bank official and travelled a lot, so I had to take care of my two children also. I would delegate work to various people in the office till seven or eight pm, so that it would go on even in my absence. Then I would drive home and see to my children's homework and meals. We would dine together, I would instruct them to read and sleep. And I'd drive back to the office relaxed and work with renewed energy till one am. Next morning, I'd drop off the children to school and be the first to begin work.'

How did her family deal with the erratic timings?

If her children were awake on the days she got home early, she would explain the day's challenges to them in a dramatic story. If she was ever hassled at work, they would reassure her that they could manage on their own. She describes the 'good understanding' she had with her family. 'I never compromised on my job as a mother. If there was any parent-teacher meeting or any programme in their schools, I used to be the first person sitting in the first row. I never missed anything. You need to strike that balance, otherwise if my family is not okay, what is the point of achieving all this?' Vellarmathi is firm on maintaining a work-life equilibrium. Today her children are 'extremely independent, mature individuals—useful to the family and to the country.' Vellarmathi lays great store on being 'useful'.

I enquire if taking breaks during her pregnancies affected her career growth.

'After my first child was born, I was not able to take any long breaks for quite a while, because of assignments and deadlines. That is why there is a huge gap of seven years between my children. Other than that, I rarely take leave. I was back in office promptly after the third month of my maternity leave ended.'

She does not bother too much about setting aside time to unwind with normal distractions such as television, films or music. 'I don't know if it's God's gift, but even if I sit quietly for five or ten minutes, it is equivalent to rest and recreation for me. When I reach home, I feel like doing something for my family. I love cooking and home food, so for me relaxation is sitting with the family, seeing them eat the food I've prepared for them,' she tells me.

'If I want to thank God, I walk. I'm not too spiritual. I don't disturb God too much. I treat nature as God. Today the sun, air, water—all rule us. Nature dictates our lives. I admire nature. I recollect incidents in my life and thank nature for a number of things—my birth, education and all that I have received.'

Is there any task that is unfinished for her?

'I am a satisfied woman today,' she says with a smile. 'As a mother, daughter, daughter-in-law and wife—I have done my duty. In my work also, I am happy with my service. If I could contribute in a mission that is 100 per cent useful for the nation, that would make me happier.

'Whenever I address any gathering of women or girls, I always end my talk by saying "If I am able to do this, why not you?" Perhaps, after seeing us some children may get motivated. You can read a number of things in books, but when you see a person executing it, that provides immense confidence to the mind.' Vellarmathi's dictum could not ring truer.

Anuradha T.K.

Programme Director GEOSAT, URSC, Bengaluru

'"The Eagle Has Landed"—Two Men Walk on the Moon' read the headline in *The Washington Post* on 21 July 1969. A day earlier, an estimated 530 million watched Neil Armstrong step out of the Eagle, a lunar module of NASA's Apollo 11 spacecraft and say the immortal words, 'One small step for man, a giant leap for mankind.'

The landing was broadcast live to a worldwide audience. In India, however, there was no television at the time. Anuradha T. K., currently programme director of GEOSAT (URSC, Bengaluru) was just nine years old when her parents and teachers told her about a man landing on the moon. She was utterly fascinated and wrote a poem on it in her mother tongue, Kannada. Today, almost half a century later, after some coaxing from me, she recites it again:

> O silver moon, why is your abode amidst these black clouds?
> Are you afraid that Man is going to visit you?
> Aren't you aware that one more attempt is going on for Man to come (to you)?

'A poet scientist, how wonderful! Do you still write poetry?' I ask her and she laughs.

'I do. Mostly in Kannada and only for myself. I don't publish anything. I'm not a poet. I just jot down stuff. Sometimes, something comes to mind but by the time I write it down, it's forgotten. By evening it has just evaporated.'

Anuradha, 57 years old, is an unusual woman—an intriguing blend of pragmatic and visionary beliefs. Listening to her offering words of wisdom to young girls and boys drawn to science, I find myself wishing I was beginning a new career all over again.

Those Magnificent Women and Their Flying Machines

'Science is something where new things keep coming up. You need to keep updating yourself to keep abreast of new knowledge. You cannot tell yourself that your exams are over, you have got your degree. Every day is a test. You are a student forever in this field—nobody is a master. The more you know, the more you understand, the more you get to know that you don't know anything.'

'Is this especially true of space science?' I ask.

'It's true of any science or technology field. In space, most of the real scientists—a lot of us are engineers, though our designations are those of scientists—become philosophical as they understand more and more universal truths, because space is so vast, so beautiful, so wonderfully exciting. Every day it throws new things at you. You understand the moon but you don't know it, even though it is the nearest celestial entity to you. Even today, debates go on about whether the moon is part of the earth that split into two or if it was some other body captured by the earth. Today's newspaper talks about a lost continent under the Indian Ocean. A lifetime is not enough to learn all there is to know about our own earth. It is so beautiful—just immerse yourself in that. Everything else will follow. Keep enjoying what you're doing, don't think your final exams are over. That's not the way a scientist thinks.'

Anuradha is one of ISRO's seniormost female scientists. She was the first female project director of the successful GSAT-12, GSAT-10 and GSAT-9 satellites, along with being project study director of the GSAT-17 and GSAT-18 satellites. The GSATs are high power communication satellites and are part of the Indian National Satellite system (INSAT)—one of the largest domestic communication satellite systems in the Asia-Pacific region. Today Anuradha is the programme director with nearly fifteen ongoing projects in communication, navigation and meteorology.

On 15 July 2011, when the 1,410 kg GSAT-12 was successfully launched into geosynchronous orbit at Sriharikota, Anuradha termed the feeling akin to 'delivering a baby'. Her all-woman team, comprising mission director Pramodha Hegde and operations director Anuradha Prakasham, echoed this description in press interactions.

Both Anuradha and the then-director of URSC, T.K. Alex, played down any gender distinctions. 'Until I was asked, I never thought it was a big deal. These women are outstanding scientists, and it is our priority to hand key projects to good scientists. They were not favoured due to their gender. No shortcuts were awarded to them,' said Alex. Anuradha added that she 'never thought of this as a great achievement. We joined ISRO to do precisely these things.'

Six years later, Anuradha admits that she might be a game-changer. 'We were some eight ladies who joined in 1982. At least 50 per cent of us are in very high posts now. I don't think there was ever any situation like "because you're a woman you can't be given that." Initially, we did hear some stories about groups that did not want to take women, because they thought women can't work around the clock. It is *their* experience and *their* background that made them think like that. As young girls who joined ISRO at that time, I feel we made a difference. Today everybody is confident. They don't care if you're a man or a woman,' she laughs.

Anuradha is a reassuring symbol of female empowerment today, both outside and within ISRO. 'Young girls from varying backgrounds who join our organization get some kind of motivation, that they too can reach this post. Some of them tell me openly that my being here gives them hope,' she says.

Besides being an inspiration, Anuradha also gives valuable, hands-on advice for working women. 'I know that several families remain entrenched in patriarchal beliefs about what women should and should not do. That's where the struggle lies for women. So what I advise my young ladies here is to take pride in themselves, in what they're doing. Let their family do the same. This is an organization where you can bring your family to office sometimes to show them what you do. Let them talk to your bosses. Do not work in isolation. That used to be the culture that existed earlier—women wouldn't bother to see what the man was doing in office as long as he brought in the salary. He would have his own stresses and problems at office, which he could not share with his wife. Today it is not like that. He has to share the household work and she has to share in the economy of the house by being a contributor.

'Take help,' she says vehemently. 'Whether it is from parents to take care of your children or domestic servants to do household chores, free up your time to pursue your work goals. When you come to the office you have to be as fresh as a man. He is well rested when he gets here, while you get up at four am, cook, do chores, pack lunchboxes, feel drained out by nine am. You have to do whatever you're conventionally supposed to do at home, but don't shy away from taking help. Who better than grandparents to help take care of the children? You give meaning to their lives, to your own life. It's a win-win situation.

'I always say I run a factory at home, I have employed so many people! I don't care how much money I spend on this. I am not here to make money or to be richer than others. I work here to give meaning to my life. So if I want it, I can make it happen. Every woman can, if she sets this as her priority. Sometimes these priorities get smudged due to cultural backgrounds, but women should have faith it will work out.'

Anuradha had struck a deal with her children that if they studied and did well in their exams on their own, she would take a week off when their holidays began and they would all spend family time together. 'They thought that was a great idea. If I had stayed at home, all the memories the children would have had would have been of their mother always pushing them during their exams. Besides, it gave them a sense of responsibility to be independent.'

Her husband, V. Kiran, a batchmate at Visvesvaraya Engineering College and currently the general manager, international marketing at Bharat Electronics (BHEL), shares her views. So did her in-laws, who are proud of her accomplishments.

'Everybody may not be so lucky with supportive families,' I interject.

'That is when you have to put in additional energy

to make it happen, not by cutting vegetables but by convincing them. You have to make your family understand what your priorities are. It is your life after all. If you want it to be good, it will be good.'

Anuradha shares her family history to illustrate this point. 'My father came from a highly religious family. He was a Sanskrit professor, the only educated person in his family. I haven't seen a more progressive man than him. My mother was not even matriculate. She was married off when she was just sixteen or seventeen. How did they decide to bring up their daughters in this way? Was it just luck or something they worked out? My elder sister is a doctor, the other two are electrical engineers. They all pursued what they wanted. And then after marriage, I went to a different family. Is it luck or my attitude that there too it should be like that?'

Anuradha's mother backed her decision to take up engineering, instead of following her doctor sister's route. 'My father had been transferred out and when I got top ranks in the state for both medical and engineering, he sent a telegram urging me to do medicine. I told my mother I wanted to do engineering and sent back a one-word telegram—"No." He then accepted my choice.

'At Visvesvaraya College, girls did extremely well. Most were toppers, though there were only ten girls in an engineering class of sixty. Many batchmates and friends went to the US, but I always knew I wanted a family; I wanted my children to grow up in India. In those days, we had our choice of jobs because we were the toppers. My husband joined BHEL and I joined ISRO,' she recalls.

'Space was one thing that really excited me from the beginning. I used to read many magazines related to space, even in college. Mathematics and physics were my favourite subjects. And I loved electronics in engineering—it was like reading an interesting novel, understanding how devices work.'

ISRO was Anuradha's dream job and it soon became a reality for the young electronics engineer in the early eighties. Working round the clock was routine for both men and women. They stood by each other, fulfilling responsibilities given by the seniors.

'Were family responsibilities divided as well?' I ask. 'Or was it assumed that you would get the lion's share? Do husbands ever thank their wives as much for their support in their careers and are *they* ever asked how they manage home and office?'

'Let me tell you frankly—without my support my husband could not have become a general manager at BHEL. He says as much to everyone in his office,' she says lightheartedly. 'My home is very far from the ISRO campus. In the days that I worked as a young engineer, we would all go home in shared taxis. If it was late in the evening, my husband would come to pick me up so that I wouldn't have to wait for a taxi. I would try to dissuade him but he said this way he could spend time with me. Why should I not thank him for that? I'd see my husband and kids waiting for me outside our home as soon as I finished work at eight-thirty–nine pm and it made me feel so good. I'm thankful to them for that too... In earlier times, when launches were rare, there used to be a big ceremony at the time of each launch and every project director would thank their families—insisting they were part of the project as well.'

Apart from the hectic schedules during a launch, when she was likely to be at Sriharikota or the Master Control Facility at Hassan, 180 km from Bengaluru, Anuradha usually leaves home at seven-thirty am. There is no fixed return time. 'All days are not frenetic. When I can't go back I just alert the family. Everything's taken care of at home, the house runs without me,' she says nonchalantly.

'Do your children [two daughters, both engineers] ever complain about you having missed out on anything due to your workload at ISRO?'

'I hope not,' Anuradha retorts. 'I have asked them this question several times, and they tell me they remember the wonderful holidays we had together. They are so proud of me and I, of them. They are absolutely independent girls, to the extent that they may not listen to me sometimes. But that's all right since that's the way I've brought them up.'

'Today's millennials plan their goals well in advance—their wishlist usually includes owning a house and a car by age of thirty, becoming a director in their organizations by forty, a chairman before fifty and so on. Was it like that for you?'

Anuradha grins conspiratorially. 'The moment I joined ISRO, I wanted to be the chairperson of the organization. I used to think, "What should I do to get promotions, so I can become the chairperson?" It appears so childish to me today, I can laugh at it. Though it probably helps in some ways to have a bigger goal:

you start observing how seniors are solving issues, you ask yourself what you would have done in their place. You start thinking, maybe not of becoming the chairman, but your immediate boss and the next boss. You start preparing mentally about getting there. That is why it did not come as a surprise to me when I became a project director, or later, a programme director. It seemed the natural course of action. I've tried my best to be good at whatever I do and I'm happy.'

Anuradha can name several highlights from her 34-year-long career, as she became one of the top scientists at ISRO—from test engineer to design engineer, project manager, deputy project director, project director and finally, programme director. She started out making electronic designs for testing the satellites, building features such as automation, and making her own protocols and designs to make things simpler.

'Whenever a complex problem was tackled and solved, it gave me immense joy. In my first assignment as project director of a communication satellite, GSAT-12, in 2011, there were many new concepts that were brought in. So when everything worked out as planned, it was simply great. Another time, before I became programme director I was simultaneously heading three projects—one of them was so complex that in the middle of the half-built satellite we had to include a new concept of electric propulsion [GSAT-9]. We did it in a beautiful and unique way, which gave us a lot of happiness,' she enthuses.

'How did you handle the roadblocks en route?'

'Sometimes you have to recede,' she smiles. 'When

you put your heart and soul into something, and if it doesn't work you have to revert to other ways to do it.'

Anuradha is currently working on building a new class of satellites—the I-6K bus,[46] a six-tonne satellite with very high power generation and complexities. Alongside this, she is also working on expanding the existing I-3K, so it can become a four-tonner.

I ask what drives her to keep searching for new goalposts, new inventions and new discoveries.

'As a scientist or engineer it is an innate feeling to keep making things better.'

'But how do the ideas come to you?' I press.

'All our programmes are oriented mainly towards societal applications. The satellites are built to carry these applications as payloads. The ideas for applications are unique to each country or space agency. Space giants like NASA, ESA and others have made wonderful payloads and are at the forefront of technology. However, we keep adapting the technology with innovations to suit our applications. For example, Astrosat is a beautiful observatory in the sky and people worldwide have appreciated it. Chandrayaan-1, MOM, the navigation constellation NavIC, and the improvements in remote sensing techniques are other examples of innovation.

'We want to do human space programmes, to at least put a person in space. Even though there's no approval for this as yet from the government, we keep doing experiments and equipping ourselves. One day we will get the approval too.[47] The Space Capsule Recovery Experiment was one such experiment for the human space programme. All of these programmes need a lot of technological development, which are reviewed very seriously each year,' Anuradha explains.

~

There are two Indias that exist today, and the disconnect between them often seems insurmountable. In the first we rank 131 out of 188 in the Human Development Index and 108 out of 144 in the Global Gender Gap Index. That is, a woman in India earns less than a quarter of the annual income earned by a man, while her share of unpaid household work and child care is 66 per cent—to the 12 per cent share of a man. Only 27 per cent of working age women are employed, according to World Bank data (2015–16).[48] While the wage gap between men and women has narrowed from 48 per cent in 1993–94 to 34 per cent in 2011–12, women still earn less than men. In this India, girls are routinely denied basic rights to survival, safety and education—2,39,000 girls under the age of five die each year due to neglect caused by gender discrimination.[49] They are forced to walk on a road that leads nowhere.

In the second India, girls want to become space scientists, fly on a rocket to Mars and participate in human space programmes. In a recent survey conducted by Naandi Foundation on teenage girls in India, a whopping 70 per cent girls said they wanted to pursue higher studies and had a specific career in mind. Critics often claim that the money spent on space research could instead feed and educate millions of India's impoverished masses, especially disadvantaged girls.

To this, Anuradha says, 'There are hundreds of

people who question what we are doing and the money spent. But if you don't invest today, it doesn't happen tomorrow. I try and make people understand—however poor you are, you make a sweet on a festival day because you want to make that happen. You need to bring out the best in you to do that. If you don't even attempt to do that, how will it ever be possible?'

Social applications of satellites help in bridging the gap between the two Indias. Tele-education, tele-medicine and Village Resource Centres (VRC)—applications of satellites such as GSAT-12—can effectively be utilized by every gram panchayat. This is Anuradha's dream. 'Ultimately, it all goes back to the basic level of how people live in remote villages. How can a farmer live with the same self-respect as I do, that is the question we all need to answer. Space can provide the solutions for all this.'

Over 6,500 programmes in agriculture development, fisheries, water resources, women empowerment, computer literacy and vocational training have been conducted by 461 VRCs in twenty-two states, with over five lakh people availing these services.

~

'After being publicly lauded for being the first woman project director for GSAT-12, are you looking forward to a time when the "woman" tag will be redundant?' I ask Anuradha.

'I am waiting to see a day like that. Frankly, it annoys me sometimes because we do not consider ourselves as some different species. If you have women awards we will get women awards. Don't keep separate awards for women and men, especially in science and technology. Perhaps earlier, this was needed to promote women but I think the time has come to stop this. Imagine if you had a Nobel Prize for women scientists, how silly would it be?'

Yet when it comes to the creamy layer in scientific institutions, women often do not make the cut.

'If you see in our own organization, we don't have a woman as a centre director for any major functional unit. But it's not because women cannot get there. Today at Thiruvananthapuram, Ahmedabad, Bengaluru—all major centres—you will find that those women who joined with me are all deputy directors. Perhaps if I was five or six years older, I'd have been a director. There is a growth pattern here. It's very unusual that someone jumps ahead, it is not like in a private company. In ten or fifteen years this question will not be asked. Some of the youngsters are doing exceedingly good work.

'What has changed in the last thirty years is that the value of education has increased in urban and semi-urban areas. In any middle-class family you will find parents ready to sacrifice anything to ensure their son and daughter get the best education. The parents of the girls joining ISRO today are probably my age, and understand the need for a girl to have her own identity. The numbers have thus increased along with the aspirations.'

However, the 18 per cent strength of the female scientific/technical workforce at ISRO signifies only partial gender parity. Anuradha attributes this to a larger trend of boys and girls chasing the big bucks at IT companies.

'So the passion for science is not too popular then?'

'Maybe the passion for having a good life comes before science,' she says wryly.

'Are there any self-imposed restrictions on behaviour with colleagues and subordinates at ISRO? Do women have to make any efforts to not show emotions that may translate as weakness?' I ask.

'I don't think any man is tolerated if he's throwing a tantrum. People will avoid him. Naturally, you have to keep calm in front of everyone—juniors or seniors. Women do need to take care of certain things. If you want a level playing field, you also have to invest in it. You cannot act differently.'

Her voice turns steely, hinting at the command she must wield at work. 'I don't say emotions are bad. I too got emotional and had tears in my eyes when the first PSLV was successful. We had failed before and everyone was waiting, hoping it would succeed. Today PSLV is one of our best vehicles and people have forgotten the earlier failure. But that emotion you show is due to your passion for your work.'

Do women call out any sexist bias of male colleagues with females, or is it ignored quietly?

'The male-female clash, in any part of the country, you will find that it is there. Within ISRO, the organization cloaks it. The culture of the organization ensures proper behaviour. Anyone treating women in a different way is not appreciated. Still, we do hear of some things and action is taken immediately. The management here is very cautious. Women are free to bring up sexual harassment in separate forums.

This is an organization where there is perfect understanding, be it a man or a woman. If you have a personal problem, there is help and cooperation. We often work twenty-four hours continuously, with no holiday breaks. So the management also responds in a different manner. ISRO is like a home. I often wonder if things are like this in other organizations or if we are an island.'

The world inside an ISRO centre, whether it is URSC, SAC or any other, appears to outsiders like me to be protected—untouched and unconcerned with routine societal pressures. I ask if this is an unfair perception of ISRO scientists.

'We are part of that same society too, so naturally we feel a connection. We don't live in office all the time,' says Anuradha. 'But I strongly feel you should not get distracted from what you're doing. You have to do justice to the task you have been assigned to do. I cannot be a doctor or educationist or a politician. Let schools and colleges excel in what they do, let doctors focus on their work, let ISRO concentrate on theirs—if each one simply does their own job, everybody will excel like ISRO,' she remarks. 'People have passion and goals here and it is a good organization with a clear path ahead. You know, my husband says no one should talk to me about ISRO—I cannot stop talking about ISRO.'

Chapter 5

Beyond MOM: The Applications Achievers

There is no problem in science that can be solved by a man that cannot be solved by a woman.

—Vera Rubin, pioneering astrophysicist

They have been described as the 'backbone' of the organization, the 'invisible' force doing impeccable work across myriad missions, disciplines, ages and seniority. Working in different satellite programmes—communication, navigation, remote sensing, oceanic sciences, digital payloads, design, data products and processing, measuring ice in Antarctica and a lot more—these are ISRO's women of substance.

They are also women who head departments at a young age, sit on recruitment committees, have a genuine passion for research and pride in their jobs. All of them, whether they have worked at ISRO for three or thirty years, believe in 'serving the country' through their collective efforts. This earnest declaration takes on a normal, everyday hue as the women chart new milestones in India's space journey. In this chapter, thirteen feisty scientists talk about their journeys at SAC Ahmedabad, Delhi Earth Station and the National Remote Sensing Centre at Hyderabad.

Durga Darshini

Scientist/Engineer-SE, Digital Communications Division[50]

It was the launch of the first navigational satellite in the Indian Regional Navigation Satellite System (IRNSS), 1-A, in July 2013. This was also around the time when I, an Odiya, was to get married to a Punjabi whose family home was in Haryana. So we scheduled a nine am wedding ceremony in the gurudwara, completed it in an hour, held the reception in the evening and flew back—so I could get back to work. I took only two days off for my marriage, and since my husband also works at ISRO, he understood how I felt. So did my mother-in-law, who knows our work schedules.

Again, during my pregnancy there was some urgent work to be completed, and my due date was twenty days later, so I thought I'd manage to complete it in time. But the baby decided to come earlier, and I had to rush to the hospital from my office! It was a normal delivery and my baby was healthy. I don't think my projects will accommodate a second baby though [laughs].

I work in digital sub-systems development. My major contribution is in the regional navigation satellite system, from 1-A to 1-G (a series of seven satellites for position finding), an indigenous GPS system of our own called NavIC (Navigation with Indian Constellation). Most of the sub-systems including the digital sub-system are indigenously developed and delivered. Timing is very important for navigation and position finding, therefore we did R&D, developing innovative algorithms using software and hardware to get the timing right in nanoseconds. I worked on this project from the very first day of joining ISRO in 2008 as a junior scientist, after completing my graduate and post-graduate studies in electronics. ISRO was my first job.

We enjoy working hard here because of the motivation provided by our seniors. One day we were working late into the night since the INSAT-3DR[51] project schedule was critical. Our division head told us to complete it or else go home around midnight. Three or four of us—I was the only woman among them—decided we would make sure it was completed, so we worked till 4 am. Our boss called to check on us. We told him we were just finishing the job, so he made some tea and brought it to us. It was such a caring gesture.

If you want to do something new or even in your domain area and expertise, this is the right place to do it. You get support, motivation and resources. You can think of innovative ideas for Technology Development Programmes, where we get appropriate resources to take our ideas further and demonstrate how they can be used in the future. Work is akin to god for me and innovation is my goal. I am thirty-five years old now, so I have twenty-five years of hard work ahead of me to make significant contributions to ISRO and the country.

Rashmi Sharma

Scientist/Engineer-G, Oceanic Sciences Division

At home everyone asks me, 'Don't you feel like visiting us?' They keep telling me to take a break. I tell them I feel happier thinking about science. Even when I get

Beyond MOM: The Applications Achievers

home from office I keep thinking about my projects. I disturb my team members at night, too, sometimes! I love my research—it is the best thing to happen to me. I have not taken my earned leave for the last ten years, so it always lapses. I make do with casual leave and weekends—never more than five or ten days. My parents are my neighbours and my in-laws also live in Ahmedabad, so I don't have to travel to different places. Additionally, I have an inbuilt support system. My son was virtually brought up by my mother. He is in twelfth standard now, with a passion for astrophysics and astronomy. My husband used to work in ISRO. It's where we met and got married. Later he became more inclined towards management, so he went ahead with his career and I continued at ISRO. He had an excellent work opportunity in the US but I was so reluctant to leave ISRO that he had to come back.

It's been twenty-five years now that I have been working here. I am heading a division here with nine people—three of us are women. In fact, SAC has several women heads of division—twenty division heads and one group head. This was not the case when I joined ISRO after completing my post-graduation in physics from Mumbai University. Out of the twelve scientists who were recruited that year, I was the sole woman.

I work primarily for oceanography satellites. India has a very large coastline—more than 7,000 km. A lot of the coastal population depends on the ocean for their livelihood, so it is important to know oceanic conditions for fisherfolk, for instance. Let's say I want to go for a holiday to coastal areas and want to know beforehand if the beaches are vulnerable or not. You need to take measurements for all of this. Since the ocean is so vast, you cannot possibly cover all of it so you have to rely on satellites to give you synoptic pictures. These pictures can give you information about the ocean at one glance.

I make use of the satellite data, put it in the numerical models and then give predictions about the state of the ocean for two, three days. We can make predictions for a maximum of five days. Ships can know oceanic conditions and take safety measures to avoid rough seas.

Another example is of potential fishing zones—informing fishermen about hotspots for fish to avoid random fishing without an assured catch. We have developed a methodology using satellite data and transferred that technology to the Ministry of Earth Sciences. The operational agencies are Hyderabad-based Indian National Centre for Oceanic Information Services (INCOIS). They reach out to fishermen through mobile services and other means of communication.

We also host the forecasts on our MOSDAC [Meteorological and Oceanographic Satellite Data Archival Centre] website on the ISRO portal. Anybody can visit that site. The Indian Navy is one of our important users since they have many fleet operations going on for their ships. For the last seven–eight years we have been working together, making customized models for them.

Our latest Samudra programme also identifies potential fishing zones in the seas surrounding the subcontinent, so fishermen don't waste time and fuel. This is one of the applications that my team and I are directly involved in. There are several other applications such as forestry, agriculture, ground water and weather information...

Along with oceanic modelling, I do coupled modelling with the atmosphere in which land is included. The entire climate system is a coupled one—you cannot segregate the oceans because the monsoon rain has its basic genesis in the water vapour rising from the ocean. So I am expanding my field to the coupled ocean-atmosphere-land model where we can have an integrated system in place for enhanced research.

As far as research is concerned, what we do is on par with international standards. I have a passion for research, with more than forty-five publications in reputed journals. I am not a very ambitious person. I always take short steps. I never think about where I will be five years down the line.

Maya Suryavanshi

Senior Research Fellow, ISRO

> *Every year the National Centre for Antarctic and Ocean Research (NCAOR), a research and development body functioning under the Ministry of Earth Sciences, organizes the Indian Scientific Expedition to Antarctica (ISEA). ISRO has been participating in these missions for several years. For the thirty-fifth expedition in 2015–16, the project team collected ground data to interpret the satellite images and installed instruments to measure greenhouse gases and atmospheric black carbon. Using the indigenously developed Ground Penetrating Radar (GPR) of 500 MHz, the team gathered data from*

seventeen locations on ice sheets, ice shelves and icebergs to study climate change.

I was the only woman on the thirty-fifth ISEA from 14 December 2015 to 1 April 2016. To apply for the ISEA, a senior male colleague at ISRO, Rajendra Singh, and I sent our proposal to NCAOR, explaining our purpose of going to the Antarctic. Once the proposal was accepted, we had to pass a medical fitness test conducted at AIIMS (All India Institute of Medical Sciences), Delhi, which was followed by a fitness-training regimen of twenty days in Auli, Uttarakhand.

Our journey began at the NCAOR headquarters at Goa. After completing the necessary formalities, we left for Cape Town. From Cape Town to the Antarctic, our ship voyage took around nine or ten days. They have their own terminologies for the crossing of the latitudes—we left Cape Town at around 34 degrees latitude, then crossed the Roaring Forties, the Furious Fifties and finally, the Screaming Sixties.[52] Even though I had been prepared for sea-sickness, I still had a tough time.

The Russian cargo vessel, *MV Ivan Papanin*, which was used for the voyage, had on board thirty Indian expeditioners and twenty-eight Russians to maintain the ship and supervise the journey. There were three Russian women too, but I was the sole Indian woman, and they all took care of me like a daughter. Apart from ISRO, there were researchers from the Botanical Survey of India, Wildlife Institute of India, and the National Hydrography Institute.

The two Indian stations in Antarctica—Maitri, which

Beyond MOM: The Applications Achievers

uses Greenwich Mean Time and Bharti, which is in the Indian Time Zone—are full-fledged laboratories. We visited Bharti first, staying there to take the GPR measurements and then proceeding to Maitri.

Our main objective was to study the stratigraphy of the ice.[53] The first layer is nothing but the snow form, but once the snow gets denser and denser it turns into ice. So our aim was to get the measurements with a GPR of 1 GHz—with more frequency it will have higher resolution but lesser penetration. We carried two GPRs, the other one was of 500 MHz made at SAC, ISRO.

Each time an expedition goes to the Antarctic, we carry the GPR because we have to see if it is working or not. If something goes wrong there, we don't have anything. We use the GPR data for snow depth calculations. The 1 GHz GPR cannot penetrate the entire area—the average ice thickness above land is 2.5 km. Plus if there is moisture, the GPR cannot penetrate. So we take readings and measure 2–3 m of snow thickness and sea-ice.[54] The thickness of sea-ice in Antarctica is 1.5–2 m.

In the thirty-fifth ISEA, we had to cover GPR points in various land forms as well, so we used helicopters on board the ship to cover these locations. Had they malfunctioned for any reason, we would have been unable to reach those locations. Blizzards in Antarctica are very unpredictable too. Fortunately, the ones we faced were not harmful. Antarctic summers are now getting warmer, period-wise as well as temperature-wise. We are analyzing the data and seeing the consequences.

I am very happy I got this opportunity. It was

something I always wanted to do. I had applied for the ISRO position of junior fellow, and when I got the interview call I asked my father several times if it had come to me by mistake. Since I have an MSc Physics from Mumbai University, I was asked about the basics of classical mechanics and Keppler's laws. I managed to answer and got through. I didn't know where I would be placed—it was sheer chance that during my meeting with the deputy director, people were discussing Antarctica. That is basically my study region now—research-related satellite data analysis in low temperature regions like the Antarctic. I am now studying energy and mass balance of the Antarctic ice shells, using remote sensing data, for my PhD in Gujarat University. Now others will get the chance of going to the Antarctic. I have had my turn.

Shilpi Soni

Scientist/Engineer-SF, Division Head,
Microwave Tube Development Division

I am from Sagar, a small place in Madhya Pradesh, best known for its vicinity to Bhopal. I was always a studious child interested in science and mathematics. I would read books rather than watch TV. In one of my scrapbooks, I remember I had written somewhere that I want to be an ISRO scientist. We would sleep on the terrace in summer and the stars would fascinate me. At the time I didn't know that was a different domain—astrophysics. I only knew that if it's space, it must be ISRO.

After my twelfth standard my father encouraged me to do engineering but since I was the first child among

Beyond MOM: The Applications Achievers

four daughters, it was a new experience for him. So he suggested the engineering college in my hometown. I might have had better chances elsewhere, but I may still have ended up at ISRO. I am currently at India's premier institution and that's all that matters.

The common man thinks space science is a hi-fi kind of a domain, far removed from him. Actually, ISRO works on many applications for the average Indian citizen, such as cyclone warnings, disaster management systems, storm and distress alerts for fishermen and lots more.

We work on forecasting. We have separate satellites such as INSAT-3D and INSAT-3DR for distress alert application. We also work on unmanned railway crossing management. Many accidents take place at such places, since people do not follow written rules. But they can follow sirens and alarms. So we are working on a hooter alarm that switches on when a train is two km ahead of the crossing.

I have worked on the latest technology right from the time I joined ISRO sixteen years ago. For the domain in which we work, the learning never stops. We are always in search mode. We see what others are doing and ask ourselves if we can do better.

My husband also works for ISRO, so he understands the environment. This may not be the case for others who face family constraints. You need to work harder because you have multiple domains to handle. Technically, you have to work the same as your male colleagues.

ISRO women scientists are flying high right now, the ratio is slowly increasing... There aren't too many women scientists at the top. This is the only concern. Though this is also the problem faced by an earlier generation. For our generation, the numbers are probably enough. By the time I reach that age, the ratio might be 50 per cent.

When I was the division head for payload filters three years ago, some of my colleagues in the division were older than me. Some were more senior as well. I had to convince them tactfully to cooperate with me. Sometimes I took advice from my mentors about how they tackled such issues. When I was given this position of division head, I called my husband and told him I was too junior, and that I wouldn't be able to cope with the responsibilities. My child too was only around five years old at the time. But my husband was sure of me, even when I wasn't sure of myself. He told me to go ahead and today hopefully I have proved him right.

Aasiyabanu N. Topiwala

Scientist/Engineer-SF, Digital Communications Division

My father was a scientist at ISRO before I was born, so people used to joke, 'Tumhari ragon mein hi ISRO hai [ISRO is in your blood]'. I grew up listening to stories about Vikram Sarabhai. I already had a passion for science, since both my parents were science graduates. I wasn't really inclined towards working at ISRO, but mine was somewhat of a *Dangal* story.

I belong to a traditional Muslim community and we lived in an area where girls were not too educated. In the eighth or ninth standard itself, girls in our neighbourhood would get engaged and after the twelfth standard or first year of college they'd get married. I was very good in studies and my father was determined I should study further. So I took admission in electronics engineering at a college in Nadiad—50 km from where we lived in Ahmedabad—because I didn't qualify in my home city.

I would travel by train daily. It was quite a struggle, since when I'd come home from college everybody would stare at me and pass comments like 'Why isn't she getting married or at least engaged?' They would pressurize my parents as well. I'd have to listen to comments about my complexion turning darker with all the train travel, my hair losing lustre... Some social barriers had to be overcome.

Interestingly though, when I used to see my childhood friends all married, with their own homes and children, I'd envy them their comfortable life. I would question my own life and what I was doing. I understood the importance of being independent only later on. My mother was also independent, and even though we never had a money problem, she would take stitching orders from people as a hobby so she could be self-reliant. My siblings are electronics engineers as well and well settled.

I completed my graduation and then the search for a boy began. In our community the boy's side contacts the girl's parents, so a lot of proposals came in. My father wanted an educated groom for me. Somehow it all clicked and I got a doctor husband, followed by a campus internship for six months at SAC, ISRO, thanks to my father.

ISRO was never my first choice though. I thought of it as 'Ghar ki murgi, dal barabar', but my father made me apply for it. At that time in 2000, there was the Y2K boom. Everyone applied for jobs in the US, as did I, even though I was apprehensive about my social background. I got three job offers, but by that time I had got engaged and my husband wanted to work in Ahmedabad with his parents, who are doctors too. He made me stay back in India. The interview call from ISRO came at the same time. I still hankered after the US job. I even completed all the visa formalities, hoping I'd be able to go some day.

But I joined ISRO and since it was a very new kind of a job, I liked the research work. So then, I rationalized that perhaps I would work here for four or five years, gain some valuable experience and *then* move to the US.

I was young and enthusiastic, and received plenty of encouragement from my bosses. I started enjoying my work. Gradually, I relinquished my US dream but I would still like to visit space agencies like NASA, ESA and JAXA...

Today I feel happy and comfortable at ISRO. There are no barriers or limitations. You can realize your dreams here. If I want to do research or development in some field, nobody stops me. For instance, I put forth one Technology Development Programme (TDP), which hosted a payload for satellite communication for disaster management. This type of system helped in the Uttarakhand earthquake, powering satellite mobile phones for people in risk areas.

I design digital payloads. I make the hardware as

Beyond MOM: The Applications Achievers

well as software for digital sub-systems, which was quite a learning curve with the shift from analog. Now we fly digital sub-systems in the payloads, reducing the bandwidth utilization. ISRO's work environment is pretty gender inclusive. If you can convince your boss about your objectives and goals, he will allow you to go ahead with your R&D, even if you want to buy things for your research. No one says you are wasting money. Sometimes you do face chauvinistic or condescending remarks from seniors or subordinates. You treat it as part of the game. Such things can happen anywhere. It is up to you to set your limit and then start countering, either with a smile or with a firing.

I do feel it is very difficult for a woman to get to the top at ISRO and become a director. A male mindset assumes that a woman may not be able to finish a project, so why give her a new one? Even maternity leave sometimes creates a negative impression, because you are out of the job for six months and you have to work harder to climb back on. That was the reality in 2004, when I was pregnant. Today girls are more ambitious—they ask for flexibility to work from home, instead of six months leave. And though parents and in-laws may be supportive, it is the mother who bears all responsibility. We are built like that hormonally. How can you escape and try to be in a man's shoes? Why would you *want* to do that?

I think ten or fifteen years down the line, we will see many more women scientists in India. The flair for science is increasing. More and more people are keen to learn. Earlier space was understood only as the moon and stars for most people. They did not know about satellites moving around the earth. I have a model of a rocket in my home. Visitors invariably ask, 'What does a rocket do? Does it reach the stars?' Slowly, the knowledge base is increasing.

I once visited a municipality school to give a talk about what we do at ISRO. I spoke about something beyond the stars, beyond planets that we cannot see. The children—from the seventh and eighth standard—were very inquisitive and eager to know about space. I don't get too much time to do this. But I am planning to, after I am sixty years old and retired.

Harshita Tolani

Scientist/Engineer-SE, Project Manager,
Radar Development

When I was selected for ISRO's Young Scientist award and the Indian National Academy of Engineering's Young Engineer award in 2016, my father-in-law wanted to announce it in the newspapers—that's how proud he was of me. He told me that the award comes with many responsibilities and I shouldn't ever forget that. I was so touched. What more could I expect? No matter how much you think you are self-sufficient, you are not able to work and excel without family support. Sometimes your project requires you to be in office at three am. My husband, Abhishek, who also has a very demanding schedule at ISRO, is very considerate. In fact, when the award ceremony took place I was away at NASA on assignment, so he collected it on my behalf.

Beyond MOM: The Applications Achievers

My first NASA visit was a great experience. I am working on the development of active antennae for the NASA-ISRO SAR mission (NISAR) as the focal person coordinating with NASA for systems of the active antennae payload. We visited the Jet Propulsion Laboratory for some testing and they were very warm, helping us with whatever we needed. They come here on frequent visits. We have a good network with them. I wouldn't want to go and settle there though—I can never think of leaving ISRO.

I chose ISRO over Infosys. Infosys recruited me on campus, but I applied for ISRO as well. I quit Infosys in one and a half months, joined ISRO and since then it has been phenomenal. The remuneration may not be as high, but the work satisfaction you get here is not possible in private companies. ISRO also has a very good reputation worldwide, so my friends who work in private companies feel proud they know someone associated with such an organization. They never show off that they earn more than me.

My professional role here is electronics development for various microwave sensors, mostly for climate monitoring, weather prediction and disaster management. I have worked on Oceansat-2, Scatsat-1 and right now, I am working on the Chandrayaan-2 lander for the unmanned mission to the moon.

I work as a project manager for radar development, leading a team for the development of very small receivers that can fit into your mobile, which would actually replace GPS. We are currently in the process of testing a few versions. Then we will put it into production, get it to the market and ultimately to customers.

My most significant project involved working on the development of the scatterometer for Oceansat-2.[55] I put my heart into it. It took about a year for us to complete activities for flight mode because there was a quality check at each step to ensure it worked for the designated five years. It became a payload that brought worldwide recognition to ISRO. I was away at IIT-Kharagpur doing my MTech when it was launched in 2009. My group director helped me complete my MTech as a sponsored candidate—the first year at IIT-Kharagpur and the second at Munich University in Germany. ISRO facilitated this just two and a half years after I began work here.

Women make up one-third of our division. Our director has announced he wants more women in upper management, so we will work towards that.

Jalpa Modi

Scientist/Engineer-SG, Microwave Data Processing Division/Signal and Image Processing Group

I never told anyone at ISRO that I need a ramp—I am physically handicapped after getting polio as a six-month-old baby—but people around could sense it and it was automatically done. My husband works in a private organization. He doesn't get this kind of support. I am treated differently in different places due to my handicap, but here I get an equal platform. I am not stopped from doing anything I want to do. Elsewhere people are not so intellectual or understanding. In a bank, for example, you will be given a clerk's job, stuck with monotonous

Beyond MOM: The Applications Achievers

work with no flexibility. I am not blaming banks—my father was a banker—but the mindset is different.

ISRO is heaven for me—I simply work here from my room, travelling everywhere with my mind. ISRO has centres across the country. Being handicapped is not an obstacle at all. I also don't have to bother about what gender I am interacting with, which is great. I'm a person who likes to do new things every time. This organization gives you many opportunities for that, along with flexible structures, so we can work independently as well as in a team.

We are called the 'data products people'. For example, we have a sensor, which measures the data and gives it back to you. But these are just numbers, which one has to interpret for the users. This is my job here at SAC. Whatever sensors we get, we convert them and make them usable for the general public.

Earth observation sensors are of two types—optical and microwave. Optical sensors take images from very high altitudes, which can be used for surveys, crop estimation and change monitoring. I work with the characterization of the optical sensor, to know how it will behave and what it will do when you fly it. The second kind is the microwave sensor, which can penetrate clouds so that we can monitor the earth round the clock, 365 days. These sensors are for climate observation.

I've also worked for the Oceansat-2 scatterometer (Oscat) and then for Scatsat-1—both sensors aimed at deriving wind velocity from the surface of the ocean.[56] Once you predict wind velocity you can track upcoming cyclones and their paths, estimate the devastation they may cause and alert people accordingly. With Scatsat-1 we could improve the sensor for land applications, use scatterometer data to detect changes in ice, observe urbanization, see crop estimation and forecast a forest.

For the Oceansat-2 scatterometer, we used to give results at 50 km resolution data, meaning every 50 sq km you have one observation. With R&D and ISRO flexibility, in Scatsat-1 we are now able to give data every 2 sq km. Achieving this feat was a career high for me.

Today I feel I am getting more than I asked for and I should not complain about what I did *not* get. There are hurdles that one learns to handle. During Oscat, we were working on a difficult problem that took more than a year to solve. At that time, my baby was six months old and I had to come in to work every weekend and stay back late many times. I was very frustrated and asked my boss, 'Is this ever going to end?' Nobody ever knows when a problem will be solved. During my maternity leave also, I missed out on some exciting work my boss wanted me to do. My promotion was delayed by six months and I felt very bad. Today, my daughter is seven and it seems like a negligible issue.

I would like to advise young girls studying science to do two things: first, get your fundamentals clear in your education. The problems you need to overcome here have no textbook solutions. Second, make sure your life partner gives you wings to pursue your dreams.

I am lucky my husband and family understand my priorities. That is why I can work as well as a man. I don't have to worry about what is to be cooked in the evening. My husband is an equal partner. He can also manage our home well.

During launch times, nobody has their head in place. Everybody is mad and working 24x7. At those times you cannot say, 'I have a daughter waiting at home'. Of course, when I spend a day with my daughter I feel good. For me, this is the best relaxation method to beat any stress.

Aarti Sarkar

Scientist/Engineer-SG, Electro Optical Payloads Integration Division

I work with optical payloads from the design stage to realizing the payload and ultimately sending it off to URSC Bengaluru for integration into the satellite. It is like sending off your daughter after marriage. You should be satisfied that you have taken good care of her and brought her up nicely. And when we get the first images of the payload from the satellite and everyone praises them, there is no better feeling in the world.

I enjoy meeting fresh challenges in my work, finding out the causes of new problems and then solving them. I find I perform even better if I'm stressed, have a time constraint and have to deliver without compromising on quality. In the final phase of our work, it's all about management of time and people to ensure that the payload/satellite performs well in space.

ISRO was my first job more than twenty-five years ago. I am from Dehradun. I graduated from IIT Delhi, followed by an MSc in Physics from IIT Roorkee. Whatever I studied got directly applied here, so I was very happy. There were very few women in ISRO at that time but there were no restrictions saying that I should not handle mechanical stuff, or fears that I would go home at five-thirty pm. We were expected to work if we wanted to rise. At best, I would get dropped off home first in car pools with colleagues during night shifts.

I was given responsible posts in the administration as well as the projects. I am deputy project director for the Cartosat-2S series along with its successor, the Cartosat-3.

I serve on recruitment committees where I see girls performing better than boys. Sometimes I joke that you may need 'beta padhao' [teach your sons] slogans in future. I am also the chairperson of the sexual harassment committee here, not merely as a token woman but as an active contributor. All the important decision-making committees have one or two women scientists. We are waiting for a time when there will be equal representation. In our own office, there is a 60:40 male: female ratio, which is increasing daily.

Another unique aspect in ISRO are the husband-wife teams. There are at least forty couples working in SAC itself. My husband was the associate project director for the Mars mission. However, despite the growing gender parity, only women are expected to do the multitasking, irrespective of how good they are or at what level they are in their profession. Yesterday I was cleaning out cupboards in my home. I put on some music and enjoyed the task, instead of fuming about why I should clean my husband's cupboard...

My daughter is grown up now, studying engineering and living in a hostel. But when she was younger, every day around six–six-thirty pm, I would get the feeling that I should go home and attend to her. Now I am free. Work gives me a lot of happiness, new goals, new problems and a great work environment—all contributing to good results.

Sometimes, of course, you face failures also. But that becomes a learning experience. Today we have reached the stage where we are confident of delivering each payload within just three months. It used to take us two and a half years earlier.

Shruti Sinha

Scientist/Engineer-SF, Microwave Sensors Receivers Division/Microwave Remote Sensing

One day I want to fly in space as an astronaut, feel what it's like to go against gravity, with people floating around the way you see in the movies. We have a unique space research programme. Once we realize the human space programme, only then can we move forward. The module that is being planned is for a human being to orbit in space up to 1,600 km, in eleven rounds, and return to earth within seven days. They will require physically fit people, which is par for the course for astronauts. To go into space, that is my dream. Though when I joined ISRO in 2005, I didn't want to be an astronaut.

I was a topper in electronics and communication at my engineering college in Rajasthan University. My friends and I were all looking for secure jobs, applying in different companies. Some got into Wipro, some

Infosys. I got selected for ISRO. Once you're in ISRO, you don't ever want to get out. Now every time we launch a satellite, whether I am linked to it or not, I get congratulatory messages from my friends at Infosys and other companies. I revel in the pride. People keep asking me 'What is ISRO doing next?'

I work on microwave remote sensing which, simply put, means you remotely try to sense something without touching it. If you try to do it from here, you will have a limited view, like that of a camera. That is why you try to look from space, where you have a holistic view of India, or even the entire earth. Now there are two ways of doing this—by optical camera use or by microwave remote sensing. In optical satellites, if there is cloud coverage the camera cannot view it. That is why you need microwave remote sensing, where we actually send waves and receive them. Depending on how they are reflected, we try to sense what is beneath them.

One of our projects is to give weather predictions for cyclones, thereby helping save lives. Our Oscat was an excellent instrument along with the scatterometer on the follow up Scatsat-1 mission in 2016. Global space agencies such as NASA and JAXA use our data to predict hurricanes, typhoons and cyclones.

The work environment here is very good. It becomes your passion. We come in on Saturdays and Sundays out of choice, not by compulsion. We often work late nights, go home for dinner nearby, in the ISRO colony just 1.5 km away and come back. In a private company, people may feel resentful about returning to office but we *want* to get back to our projects. My friends working in the private sector deal with the competitive syndrome all the time, but in ISRO, it is never a one-man show. You need teamwork to succeed.

As for me, I follow Steve Jobs' advice—there is one life, you have to live it right now. Aaj karna hai toh aaj hi karna hai. Kal nahi dekhna [Complete your work today, don't put it off till tomorrow].

Deepti Patel

Scientist/Engineer-SF, Digital Communications Division

> *Just four days before Cyclone Vardah hit Chennai in December 2016 the India Meteorological Department started tracking it through the images beamed by ISRO satellites, INSAT-3DR and SCATSAT-1, orbiting 36,000 km above the Earth. These satellites acted as sentinels in the sky, informing us about the cyclone's intensity and where it was heading. Alerts could thus be sent all along the Tamil Nadu coast to evacuate people in advance.*

I work as the deputy project director for INSAT-3DR, a state-of-the-art communication and advanced weather meteorological satellite launched in September 2016. Just a few months later it helped predict the path and intensity of Cyclone Vardah, thereby alerting authorities to take preventive measures.[57]

Along with SCATSAT-1, INSAT-3DR provided the images that you see on national news. It has daily applicability. The IMD, Civil Water Works and the Narmada project—all use its data. I have been working for payload systems since 1986, when I first joined ISRO. At any given point, I always have three or four projects running simultaneously. I am the project manager for the IRNSS navigation satellites (NavIC), designing modulators that have flown in all the seven IRNSS satellites. More and more resolution is the topmost requirement and with the new satellite too, we will give higher resolution images. We are making important sub-systems for this. It is work that requires accuracy and precision.

In case of failures, our seniors first provide moral support, and we discuss it at length. Only then are we reprimanded. We have to keep calm despite high-pressure deadlines because only then can we be creative.

I always believe—I have read this somewhere—that no one manufactures a lock without a key, and god gives no problem without a solution. I reminded myself of this while working against a severe time crunch as a new deputy project director handling many new sub-systems. I rarely face any issues with male subordinates reporting to me, but if I do, I tackle it with a smile, without anger. This dilutes matters.

Arundhati Misra

Scientist/Engineer-G, Group Director, Advanced Microwave and Hyperspectral Techniques Development

Today ISRO has many couples working in the same centre but in my time, a good thirty years ago, there were hardly any husband-wife teams. My husband Tapan [Tapan Misra, Director, SAC) and I would make sure that the office and home were kept separate. I remember, once I was doing the data processing for a project and the director was supposed to see the data at seven pm. Tapan was the systems person and was also needed in office, so I asked if I could bring my two-year-old son to work. The entire lab looked after my baby, and the director said, 'Great, the entire family is doing something here.'

I came to ISRO as a very young electronics and telecommunication engineer (later, I did an MTech in computer science). I had wanted to look for a job in the IT companies in Bengaluru, but my husband had already joined ISRO and told me to give it a try. I was placed in the signal and image processing area. It was quite unnerving for a young person with no experience to tackle a totally new topic from the beginning.

I also learnt of a new term—synthetic aperture radar processing [SAR], which is basically a radar mode that can penetrate clouds and fog. It is an all-weather imaging service that provides images day and night, and is especially useful for farmers during monsoons when clouds obstruct images. SAR was used in RISAT-1, the radar imaging satellite.

This is how it all works in a nutshell—all payloads (communication, radar, optical) are made at SAC Ahmedabad. A payload is like the eye that senses, it is the sensor. It is integrated in a satellite which is made at URSC, Bengaluru. After that, it is sent to SHAR, Sriharikota where it is integrated in the launch vehicle or the rocket. The launch vehicles come from the Vikram Sarabhai Space Centre (VSSC), Thiruvananthapuram. The images we get from the satellite are processed at the National Remote Sensing Centre (NRSC) in Hyderabad and then the products are operated by the end-users.[58]

Every week we would do innovations. In 1988–89 for example, we used to take ten hours to process an image, while the aim was to process it in as close to real-time as possible. I developed and designed an algorithm so that processing time came down to 1.5 hours, and nowadays with improved systems it takes maybe thirty to forty minutes.

Today I am handling a totally different area after putting in twenty years of learning, effort and expertise in signal and image processing. I have been assigned the role of new technique developments in the applications area, run mostly by scientists in the fields of geology, agriculture and physics. I took up the challenge to do new things, but the promotions in my career were obviously reduced since I had to start from scratch.

I have faced numerous issues on other fronts as well. Workwise, I always did well. The challenge was in getting adequate support in terms of infrastructure and backing on both fronts—office and home.

My husband Tapan went to Germany when my son was just six months old. I didn't know how to drive a car or even a scooter. My father came over from Calcutta to help out. He felt his daughter needed it, and her career and child were both important.

There were no creche facilities in ISRO in the 1990s, so I led a two-woman taskforce to install one. We approached Dr Deepti Rastogi, the only woman deputy director in SAC till date. She did pioneering work on INSAT-1A but today, not many people know about her.

It took some effort to convince her that we would work better with a crèche on campus and that we would not be taking time off to check on our children instead of working. She finally agreed to a six-month trial. We had to make do with a driver's quarters and a little plot of land. We made swings out of car tyres and that was the first creche at ISRO. I put my son in it to lead by example, and once it was successful, it was upheld as a case study in management. My son is an IIT-Kanpur electronics engineer today with his own start-up.

When I first started working here, the atmosphere was very good. Nobody treated women differently. We were all just scientists and engineers. If anyone had a problem, the entire community helped out. There was total trust. In any case, we are all workaholics spending more than twelve hours in office. I don't even go to the canteen. If I am tired, I have a cup of tea at my desk. My leave, both paid and casual, lapses very often.

As far as gender discrimination goes… I feel men can get away with making arrogant remarks, with cribbing or complaining. But if a woman behaves the same way, she will be labelled as not being strong enough. I once received some really stupid jokes and teasing comments in mails and WhatsApp forwards from senior men in Bengaluru—all sexist remarks about wives (they forget the same women are also mothers and daughters). I asked them to stop sending me junk. I took a stand and it stopped. If it had been two women doing the same thing, people would have said, 'See they don't have work so they are sending messages.'

I once selected girls among MTech students for recruitment and some boys didn't like it and accused me of gender bias. But the girls *were* better. Other times men are protective and they feel women should not be sent on field tours where there are hardships. So you just have to find a way to set new examples. Once some calibration studies needed to be done in the Rann of Kutch and women were not allowed to go. So I sent two of them there to back each other up—and what a standard they set.

So one should not ever say something is not possible. A man can get away with being only a technologically good scientist. A woman has to be a full square—everything has to be good.

Shahana Khaleel

Scientist/Engineer-SG, Delhi Earth Station

I head a team of fifteen scientists here. I have a vision of making this a big ISRO centre. We have worked hard to renovate and modernize this place. I want this to be a viable technical cell. We keep ourselves ready so that when the satellite comes we can show that these are the applications, which can be used for other satellites. For the users also it is helpful so that they can present their demands.

I will explain how our work is applicable to the common person—suppose there is a disaster, all terrestrial communication will fail, nothing else will work but satellites will always work. I can provide a terminal which can communicate and I can have a similar terminal on the other side in the district magistrate's office, or district headquarters, and effective communication can take place. So this is the way ISRO plays a role in disaster management and communication. We also have smaller phones that can be used to talk to each other, some video phones. We have terminals through which we can track vehicles.

My work is rewarding and I have received due recognition for it too. Earlier I worked at URSC, Bengaluru, for almost ten years. Perhaps I might have had more opportunities there such as the Mars mission, RISAT or other missions, but my family is here with me. That is also a source of happiness.

My husband is an engineer posted at NTPC (National Thermal Power Corporation), busy with 24-hour operations. I had to shift to Delhi after we tried out all permutations. I did not want to leave my job. He couldn't leave his either, so my bosses understood when I approached them with a request for a transfer. I had to give up the work I had done for ten years and switch from remote sensing to communication—a totally different field. I had to start from scratch, but of course, experience always stands you in good stead. I completed an MTech in computer science through part-time evening classes after four tough years, just to make myself fit in here.

My husband has his work pressures, I have mine. We are like two railway tracks, running parallel to each other. When the pressures overlap, either I give in or he does. I told him you accept me the way I am and I will do the same.

I am a very persistent person; I do not give up easily. I fought against all odds to get here. I come from a conservative background and had to convince my parents to let me do science. They supported me, even though I was among the earliest ones to be educated in my paternal family. They were all business people, so I had to persuade them about my passion for studies. People would tell my parents, 'What is the point of educating your daughter, she will get married soon and then you'll have to find a groom who is equally educated. People don't like educated women since they tend to make their own decisions.'

And when I got the ISRO job, they'd say my parents were living off their daughter's earnings, which was not true at all. Thankfully, my parents did not bother about all this and were happy with my progress. The change in mindset in our society has to come from our homes—don't encourage the son to play and force the daughter to cook, share chores equally, treat them both the same. I always thought differently from other girls. I did not want to do what they did. I was expected to do household work but I did outdoor activities as well. In school, taking part in extracurricular events helped foster resilience.

I actually wanted to do medicine. It did not happen because I was told it's not a profession for girls. As with my work ambitions later on, I always had a plan B ready, and a plan C too for that matter.

ISRO was my first job. It was very close to my heart. I began as a system engineer and worked on different satellites. I remember in those days we would be in and out of meetings and discussions. I would run to catch my vehicle in the evening, suddenly realizing it's time to go home. The atmosphere was so captivating and enlightening, each and every day was an education in itself. In those days, there were few women scientists... We used to be so busy planning how to get things done. Over lunch, over tea, we would have animated discussions all the time.

People say there are not enough women scientists in top positions here, but I think a woman can become chairperson of ISRO one day. I believe if you want to do something you are not able to do, you have to find other ways. Suppose someone asks me to wear purdah. I will do that and move on... Don't limit yourself and don't give up...

I believe having a mentor system helps a lot. I used to call all my seniors my mentors. I would catch them at the end of their work day for advice and guidance, and clear my doubts and problems.

Today ISRO has come into the limelight. People are getting to know about scientists through the internet. The economy also has a role to play as government jobs are lucrative now. The springboard is ready for people to step on and jump. Even though I did not work on the Mars mission I still get inspiration from it—that feeling that if my colleagues can do this, so can I.

For me, there is a deep sense of pride in contributing to society, even if it's just a drop. There's an intellectual satisfaction with what one has done. I am content that I made the right choice. The second best—for me—has been the best.

Mangala Mani

Scientist/Engineer-SF, National Remote Sensing Center (NRSC), Hyderabad

> It is called 'overwintering'—lasting out an entire winter in the world's coldest region, Antarctica. Mangala Mani celebrated her fifty-sixth birthday in Antarctica. She was the lone woman among a team of twenty-two men on the thirty-sixth Indian Scientific Expedition to Antarctica. She spent an astounding 403 days at India's Bharati research station.

Trusting in Lord Jesus Christ, I dared to venture into the coldest, windiest, driest and most isolated place on earth—Antarctica. A challenging and adventurous spirit imbibed by my parents made me sign on for this. I had never even seen snow before. The moment I set foot on the ice at the Novo airbase near the Indian Maitri station, I was awestruck by the serenity of this mostly untouched part of our planet. Words cannot describe the breathtaking beauty and vastness of the white blanket that surrounds you as far as you can see—calm, pure and stunning. Our task at Bharati was to operate and maintain the ground station where ten out of fourteen orbits [of passing satellites] are visible, unlike in India where only two or three orbits can be seen. We ensured fault-free, round-the-clock operations of systems so that the satellite data collected could be transferred to India for processing and distribution to users.

To make sure this operation was seamless, we worked in eight-hour shifts on rotation for the entire duration, putting in extra hours whenever required during the launch and initial phases. Apart from our official tasks we also performed galley duties that were part of station vigil/maintenance, where we would help the chef in the kitchen, keep watch through the night and take readings of vital parameters for a safe station. We did this in turn every ten days.

Initially I would wonder how people found their way here without getting lost, since there were no landmarks in the icescape all around us. A GPS device and an Iridium phone helped us keep in touch, as did the walkie-talkie radio set.

We had brought provisions for the whole year along with fuel to supply power and heat to the station—rice, dal, cereals, frozen fish, mutton and chicken, fresh fruits, eggs, milk and vegetables. In fact, whatever we eat on the mainland was available in the Antarctic, except that fruits and vegetables would last two or three months while milk and eggs were good for six months—after which we switched to powder, condensed milk and tinned fruits. Judicious use of pickles, bread, butter, jam, oats, all properly preserved, saw us though the long period.

My Christian upbringing and values helped me to stay motivated and strong for over a year as the sole woman in a place that is truly challenging on all fronts—climate, terrain, work, food and isolation. The men and I would make adjustments in terms of using the common spaces and facilities. Reciprocal respect resulted in a smooth ride. The work kept us busy and that was a blessing. Whenever I felt lonely, I would pray and that kept me going. If I noticed anyone looking gloomy or disturbed at the station, I would pray for them as well.

Those Magnificent Women and Their Flying Machines

I had a lot of personal time to sit on the terrace, enjoy a cup of coffee and watch God's awesome creation in front of me—the frozen sea, the trapped icebergs, the snow, wind and sunshine. On cloud-free nights, I would look out at the starry skies, the Milky Way, the Auroras or Southern Lights never seen at our latitudes in India. It was an experience of a lifetime and an opportunity I am extremely grateful for. I had such mixed feelings on the day of our return, having to leave such serene beauty, but also the anticipated joy of reuniting with my family. Being connected to them via WhatsApp, courtesy the satellite link at Bharti station wasn't quite the same. Taking over from me, NRSC's Ankitha Reddy is currently overwintering in Antarctica, with a long queue of young women lining up to follow our footprints in the snow.

Chapter 6

Crossing the Rubicon

I have often claimed that I have had but one good idea in my life: that true development is the development of women and men.

—Vikram Sarabhai, founder of ISRO

24 September 2014: 6.56 am to 7.41 am IST. It is the crucial forty-five minutes before the Mars Moment of Insertion (MOI). ISRO's Facebook handle for the Mars Orbiter Mission had asked its followers across India and the world to 'Be with MOM as it enters the Martian orbit,' while providing a real-time countdown to the thrilling finale. The Mars Orbiter Spacecraft is made to slowly orient its LAM engine and eight thruster engines in the requisite direction. Its Forward Rotation started at 6.56 am, decreasing its speed and allowing it to be pulled by the weaker gravity of Mars, into an elliptical orbit around it. The burn or engine firing begins at 7.17 am and continues for 1388.67 seconds, or 23.14 minutes.

As millions watch the live coverage on TV screens, the ISRO website and social media webcasts, the core team at MOX 2 track the spacecraft's progress on their monitors in hushed silence. The scientists' outward demeanour remains unaffected by any nerves they might be feeling.

The presence of the official Doordarshan videographer, documenting MOM's progress, is the first visible indication of the day's significance. A huge conglomeration of world media waits it out in an adjacent room. The Prime Minister is seated in the VIP gallery along with several dignitaries.

In just a few moments, the perfectly orchestrated efforts of 500 scientists across ten ISRO centres, over a period of eighteen months, reaches a crescendo.

~

Deputy operations directors, Nandini Harinath and Ritu Karidhal, recall their most cherished memories in the run-up to and in the aftermath of this moment.

'All the low-earth and interplanetary missions are tracked at ISTRAC, where there is a big control room full of terminals. Every terminal has an expert in front of it. The mission team—all of us in operations—sits in the middle circle, with assistants all around informing us about what is happening,' explains Nandini.

'There were a lot of long discussions and reviews on the previous day, with last-minute verifications and checking of uplinks done two weeks prior to the actual insertion,' she adds. 'Our current chairman, Dr Kiran Kumar, who was then the SAC director, was with us 24x7. When he starts a review, you're going to be there in that conference room, forgetting your lunch and dinner, till late night,' laughs Nandini. 'Drivers hired during launch times would wake us up on reaching home at two or three am. We would be fast asleep in the regular commute from Peenya.

'23 September was spent in last-minute rehearsals on what you should do and not do, and in planning and understanding what could go wrong. What if we don't get telemetry—automatic recording and transmission of data? What should we do to recover and retrieve the mission? Dr Kalam [the first mission director of SLV-3 in 1979–80] visited us in the evening, personally spending three–four minutes with each one of us. Those were precious moments.

'The reviews were conducted internally by the chairman and the project director Arunan. The checklist was ticked off. There were so many things going to happen on board which we had to clear and say, "Okay, till now we're good. We're green." At every point you look for a success-crisis criteria, and then you say you're done with this, "let's move to the next step." So it all went off well. The night was long and the Moment of Insertion was in the morning. None of us slept at all,' says the deputy director.

'At some point in the morning Prime Minister Narendra Modi came in. They had cordoned off the entire area with a lot of security. We didn't quite realize when he arrived. He was right above us on the first floor, in the glass enclosure for VIPs. And then the actual moment when the burn started, we knew it started well because we had got the telemetry for the initial few minutes. That itself was a big thing—the engine had kicked off and was doing what it was expected to do. I think all of us knew in our hearts that it would work because we had done a four-second test burn on 22 September, which gave us the confidence that despite nine months of being idle, the engine would fire and work. This was the most critical part. In none of our earlier missions had we used the engine after nine months of shutdown—the hardware is always something that holds an element of doubt. The rest of the systems—communication, navigation, sensors—were all proven systems and we had also tested the autonomy. With software, you know how things work, so checking it is easier. The satellite then took a rotation and went behind Mars.'

Ritu Karidhal explains the MOI, the manoeuvre that everyone waited for with bated breath on that September morning. 'All the planning, configurations and instructions to the onboard computers on the Mars Orbiter had been done ten days in advance. What time to start the sequence, what time to start the burn, where and how to rotate—the instructions were all loaded and verified as early as possible. You cannot wait till the last moment for these details. We had to depend on our ground stations for tracking, the Indian Deep Space Network's 32-metre antenna at Baylalu near Bengaluru, and on the other side of the earth—on NASA's Deep Space Network in Goldstone, California and in Canberra, Australia.

'We kept a watch from around thirty-six hours before the MOI—we were continuously sitting in the control centre. Reaching 500 km near Mars was assured but then the satellite had to fall into Mars' gravity and start rotating around it—this was the critically important 23-minute burn, to provide that thrust or push into the gravity well of Mars,' says Ritu.

'And then as all of us watched in total silence, the main portion of the burn was actually happening by the onboard computers behind Mars, where we could not even get any telemetry because Mars is in between.[59] We had to just watch without doing anything, but by that time we were confident that it will happen.'

~

As the minutes tick off on that September morning, the silence inside the expanse of MOX 2 is deafening. A rapt Prime Minister Modi watches the drama unfold in the VIP enclosure alongside Karnataka Governor Vajubhai Vala, Chief Minister Siddaramaiah, several state and central ministers, former ISRO Chairman U.R. Rao and former SAC Director, Professor Yashpal.

Below them, the mission team collectively stares at their monitors, unblinking, until a few seconds later the sound of clapping rises from the first row. It grows in volume and intensity till the entire complex is reverberating.

The Prime Minister declares, 'Aaj Mangal ko MOM mil gaya [Mars has welcomed MOM today]. History has been created. We have dared to reach out to the unknown. India has achieved the impossible today.'

The scientists get up from their assigned places exulting in celebration, congratulating each other, shaking hands and exchanging hugs with wide smiles on their faces. Seated next to Kiran Kumar, then-chairman K. Radhakrishnan is among the first to receive the first official confirmation from Mission Director Kesava Raju. The burn is completed at 7.41 am but the confirmation reaches MOX 2 around eight am.

Project Director, S. Arunan, is swamped by colleagues and peers on all sides, radiating pure joy—a man whose mission has been accomplished at long last.

Those Magnificent Women and Their Flying Machines

By the time the Prime Minister reaches the makeshift podium in the control room, the Mars dream team is lined up in rows to listen to an effusive 23-minute paean of praise. 'When the mission's acronym became MOM, I knew it would be successful because 'mom' never disappoints,' he says. Modi then asks schools and colleges across the country to come together for five minutes to applaud the scientists and emulate their accomplishments.

As Modi steps down from the dais, the chairman introduces the team members to him one by one, each getting a congratulatory handshake. Several young scientists request the PM to autograph the identity tags hanging around their necks and he obliges willingly, before being escorted out of the complex by the chairman.

Seetha Somasundaram, Nandini Harinath and Ritu Karidhal, their faces beaming, are among the dozen or so women scientists in the room.

~

Almost two and a half years later, when I ask the women to relive that triumphant moment, they get emotional. 'When we got that first signal, we didn't know where we were. Everyone was clapping thinking, okay, it has gone and it has reached. The satellite is now revolving around Mars. It was a historic moment for me. The experience was unbelievable—it gave us all a tremendous feeling of achievement. Even now, as I recall that time, I am getting goosebumps,' says Ritu.

'That T-0 moment is like the tying of the tali around the neck during marriage. The emotions you feel while tying the first knot, we felt the exact same emotions that day,' Minal laughs.

Nandini recounts what happened next. 'The orbit insertion was the biggest and most satisfying part of the entire Mars mission. In the celebratory aftermath, all of us jumped up from our chairs, ecstatic. For a moment, we all forgot that we were supposed to be sitting and watching the telemetry to see what happens. In that one moment, all the contingency procedures we had been discussing went into the dustbin.'

The initial excitement was followed by a huge sense of relief, and then it was back to work for Nandini's team. 'Within twenty minutes of the burn, the satellite was to take the first pictures of Mars. All of it had been pre-programmed by my team and I. While the Orbiter was going around Mars, its camera naturally happened to be looking towards Mars and we were smart enough to just turn it on with the right exposure and take a picture to prove to everyone we're there. Of course, the telemetry also indicates presence, but the picture is immediate proof. I don't think anyone has taken a picture that quickly, within a few minutes. My team had planned it all in advance—where and when to take the picture, what should be the telemetry etc. When the PM started shaking hands with people, my next worry was that now that the camera was going to start taking the pictures, have we done everything right? And then you could see, *click, click, click*—one, two, three, five pictures taken in all. We were really happy. By the time he came to meet each one of us the pictures were already taken and were downloaded an hour or two later.

Those Magnificent Women and Their Flying Machines

'That day nobody felt like going home despite being at work inside MOX 2, since the previous morning, despite being dead tired, we were in the Martian atmosphere for the first time. So we had to finish a few things—see how the satellite behaved, along with a lot of new stuff. Though we knew what the gravity would be, what the satellite's response would be, we still needed to watch it. No one wanted to leave, so the entire team was there till late evening and only then went home,' recalls Nandini.

'Once everyone left office, there were a lot of pictures and congratulatory messages. Everyone had their bit to add—the media, relatives, my parents, my brother in the US, my friends, a lot of people called to tell me they saw me on TV. We knew everyone was watching us. This was the first time something like this went live on TV. A neighbour walked in and exclaimed, "I didn't know you work for ISRO!" And then, my kids were waiting. They were up to date, following what was happening closely. The elder one kept changing her display picture on Whatsapp with comments like, "Mom finished the burn," "Mom did this, Mom did that."'

Ritu too stayed up late. 'After I reached home late in the evening, I spent time with the kids and then finally caught a few hours of sleep. My first congratulatory message always comes from my husband—I send him the timings of the launches I work on. He blocks his time ten minutes in advance and watches it online. Once it is successfully catapulted into space, he sends me the phone message immediately. We followed our personal ritual on the MOM D-Day too. And from that day onwards people have been inviting me to lecture and to celebrate. I have visited so many places.'

Nandini describes a memorable visit to Allen Academy, Kota, which coaches children for class twelve and the IIT Joint Entrance Examinations. The director of the academy showed her a video of the students watching the Mars Orbiter launch. Just after it was declared successful, each class broke into impromptu jigs. 'It made me feel so nice. First, that people watched it live and then, that they celebrated in this manner. The kind of appreciation and attention we got was something we didn't anticipate,' beams Nandini. 'Personally, I sometimes feel that maybe I was lucky to be in this project. A lot of people do good work, but you don't get acknowledgement and praise every time. This time it happened.'

~

Countless television and print interviews followed after the successful mission. The first group interview on NDTV, however, featured the likes of K. Radhakrishnan, Kiran Kumar, Mission Director Kesava Raju, S. Arunan and Anil Bhardwaj—the main men of MOM—and not the women. The women scientists were feted by various organizations across the country, though it would take a while before the ordinary Indian can put faces to the names of India's top space heroines. Even male scientists are equally unknown entities for that matter, barring the late President Abdul Kalam and ISRO founder Vikram Sarabhai. Anuradha T.K. says, 'There are so many channels on television. Which channels do people watch the most? How many viewers see programmes on the Discovery channel? Lakhs of viewers are attracted by films, sports and other entertainment. You cannot expect everyone to be interested in science and technology. Even with satellite missions, common people will follow Chandrayaan or Mangalyaan, even though we have satellites that are far more complex. Such mass appeal missions like Mangalyaan are very important to inspire children to be curious about science and to create public awareness.'

Moumita Dutta remembers feeling overwhelmed by the outpouring of greetings from people. 'I was leaving office after watching the MOI in SAC's packed auditorium. I stopped at a shop near the gate of ISRO's residential colony, and people came up to me to congratulate me. I saw widespread excitement among common people on TV, like school students holding up a banner with pictures of our project director, chairman and all the leaders of the Mars mission. I have only seen this for Hollywood or Bollywood celebrities. In the car, FM channels discussed MOM, there were children proclaiming that they too wanted to go to Mars, and schoolchildren across the country sent us congratulatory postcards. It was amazingly inspiring for the youth. I won't forget that day in my entire lifetime.'

ISRO's social media experiment proved hugely popular with over ten lakh 'likes', thousands of 'shares' and innumerable compliments and comments from enthusiastic admirers on both Facebook and Twitter. MOM contests were run with ISRO collectibles as prizes. On Facebook today, ISRO's MOM page is liked by 700,000 followers while the main ISRO page has notched up over sixteen lakh followers; ISRO's Twitter following exceeds thirteen lakh.

The next day, one picture spoke a thousand words. A bunch of women in silk saris with jasmine flowers in their hair enthusiastically shook each other's hands, radiant smiles all round. On the periphery of this photo by AFP news agency photographer Manjunath Kiran, there are a few smiling men watching their female colleagues' excitement. The caption said, 'Staff from the Indian Space Research Organisation celebrate after the Mars Orbiter Spacecraft successfully enters the Mars orbit'. The photograph went viral within minutes. The image of these struck a chord among the thousands who retweeted it with congratulatory comments. Egyptian-American activist-journalist Mona Eltahawy tweeted, 'Love this picture so much: when was the last time you saw women scientists celebrate a space mission?' Others described the women as 'role models' and 'redefining mission control'.

ISRO's seniormost woman scientist, Anuradha T.K., says that they all received it in their Whatsapp inbox. 'The motivation should come from the programmes. We are only namesakes. Today I could be programme director. It could be somebody else tomorrow. The programmes will keep running. The Mars mission is more important than the photograph.'

The Mars mission continues more than three years later, giving scientists fresh data and daily perspective. Ritu explains the more recent developments of the MOM. 'The satellite moves around Mars every three days. Sometimes it comes very near Mars, say 500 km, and sometimes it is 6–7,000 km away, so we have different modes of imaging. When you are very near, you can see with high resolution. When you are far off, you can have a bird's-eye view of the entire planet. Different cameras have been planned for use. The onboard cameras have sent more than 900 images so far. In normal times, when it is not imaging, the antennae are looking towards earth because we need to keep getting telemetry.'

On 17 January 2017, ISRO undertook a life-saving manoeuvre for the MOM—supplying energy during an eclipse—when the satellite went behind Mars and the sun was on the other side. In the absence of solar energy, the satellite can survive on batteries that can last for a maximum of 100 minutes, while an eclipse can last for more than six to eight hours. 'We changed the plane the satellite was moving on to ensure that the eclipse duration was reduced to less than 100 minutes, thus saving it from the long shadow of Mars,' Ritu explains. This manoeuvre consumed 20 kg of propellant fuel, but ISRO Chairman Kiran Kumar reassured the public that with the remaining 15 kg fuel on board, the spacecraft would survive for another five to six years.

'The shelf-life of a satellite can be more than the designed life. MOM's [shelf-life] was calculated as 180 days. It can go on for years on end,' says Ritu. Some of our satellites are surviving for fifteen or sixteen years, so MOM may well do the same.'

'Today everything is autonomous. We still get daily reports from our controllers about the satellite, we see

Crossing the Rubicon

what's happening, we review it when needed. There are people continuously monitoring the satellite.'

~

ISRO's Mission Operations Complex in Peenya is casually referred to as MOX. It is an hour-and-a-half drive from URSC, Bengaluru. The massive antenna catches the eye as one walks past impeccably manicured lawns inside the compound of ISRO's main satellite tracking and monitoring facility. It looks deserted and very few people cross my path as I walk towards the Space Operations Area. A tranquil silence envelops the entire complex.

I walk through an empty corridor into the unassuming office of R. Srinivasan, Deputy Director of ISTRAC. A tall, genial man with slightly greying hair, Srinivasan gives me a crash course on the kind of satellites tracked at Peenya, as well as the intricacies, challenges and triumphs of the Mars Mission. Only interplanetary and science missions are tracked at MOX. The navigation and communication satellites are tracked at Hassan, three hours away from Bengaluru. After completing his introductory lecture, he leads me to the main 'staging area' at MOX 2, where all the action takes place.

The space is completely awe-inspiring and it takes several moments to absorb what is in front of me. Six humongous 54-inch display screens, each showing different images and videos, immediately pull our attention towards them. Each screen displays different activities and is name-tagged with timings and locations of the ground stations abbreviated as MAU (Mauritius), BIK (Biak, Indonesia), TRL (Thiruvananthapuram) etc.

Those Magnificent Women and Their Flying Machines

The first screen displays images from Mars taken by the Orbiter, such as that of the Martian canyon, while another shows the first images of the Mumbai International Airport taken by Cartosat-2. One screen shows the now familiar, detailed Mars diagram with the process of the Mars Moment of Insertion. The last screen has a video of the actual insertion.

When there is no launch the screens are switched off. Today's videos are for my benefit, to help me get a feel of the MOM experience—without the scientists. At launch time, two screens show real-time video streams, two are used for vehicle trajectory and two for satellite health data. Each individual computer screen is configured to show real-time data and commands can only be given from the dedicated mission room.

Today, there are a couple of men sitting near a terminal in the outermost semi-circle of computer stations. The middle circle is vacant.

Srinivasan continues his briefing: 'ISTRAC currently controls twenty satellites, each having two dedicated terminals. By 2020, this is likely to go up to fifty. Our day-to-day operations are well planned. We have people working here in shifts round the clock.' The satellite orbit period is around 100 minutes but it is not visible the entire time—there are gaps. There is however an onboard recording facility of satellite health data. So whenever the satellite is visible for tracking, the data is downloaded and analyzed.

The launch simulations happen two months before a launch. All sub-systems scientists are present a week before the launch day to ensure all systems are working.

Crossing the Rubicon

They stay on for fifteen days after the launch for sub-system calibration, and once it is declared normal they hand over charge to the ISTRAC team for further monitoring.

After a refreshing tea break in the posh visitors' room with uniformed staff serving dry fruits and snacks, Srinivasan introduces me to monitoring team scientists, Roopa and Bijoy, and leaves me with them. Roopa and Bijoy are friendly and helpful, and lead me to their workspace—the adjacent Dedicated Mission Control Room (DMCR).

Inside the DMCR, a majestic bronze model of MOM with its rocket launcher occupies pride of place at the entrance. This vast room has king-sized screens of its own, monitoring different time zones and regions. As Bijoy attends to the satellite data in the form of a string of numbers on his computer monitor, Roopa gamely answers all of my queries. Since phones and dictaphones have not been allowed inside, she waits patiently as I take handwritten notes.

At ISTRAC, each scientist routinely monitors two satellites. Roopa M.V. currently has three on her watch, with more to come in the future. She is a deputy project director for Oceansat-3, operations director for Scatsat-1, Oceansat-2 and MOM. Having worked for more than twenty years at ISRO, she shares her own experiences. For ten years, she alternated between night and day shifts every week, with no holidays. 'It is certainly difficult for women. As a child my daughter would ask me daily, "Which shift today Amma?" My husband managed it all and brought up our daughter almost by himself. In those days, I used to work 24–48 hours non-stop. If there was an anomaly [in one of the satellites], I could not go home because it was my responsibility. Today it happens a lot less, perhaps once in six months or so. There is more onboard autonomy and my inputs are taken in during design time.

'For MOM they call only me, since I'm the manager. We are a team of five members monitoring the satellite every minute [they work in three shifts]. In case of an anomaly it can be corrected on the ground,' she says, showing me sixteen green markers or lines on her monitor, which indicate that MOM is working fine. 'At present, the satellite is 397 million km away. It takes us twenty-two minutes to get a signal and twenty-two minutes to react and rectify any irregularity, if necessary. But we haven't seen any major glitch since all this was taken care of prior to the launch.'

Roopa provides an example of the pre-launch diligence they undertook for MOM. 'Our general shift can last from eight-thirty am to nine-thirty pm, but before the launch there was no fixed time at all. They would send the vehicle to pick us up anytime, even at three am, if required. Extensive reviews were conducted by the then-SAC Director, Kiran Kumar, which could go on from eight-thirty am till ten pm. And then he'd say, "Can we see the results by eight am tomorrow?" People would sit up all night to deliver the results. But because he was such a taskmaster then, we don't have to stay up nights now.'

Roopa is an electronics engineering graduate from Mysore University. She completed her MTech after joining ISRO. Like the several women scientists I met in ISRO over several months, she too stresses the importance of an 'adjusting' husband to succeed at work. She also adds that taking maternity leave does affect promotion.

'For MOM, we had several women on the team—Nandini, Ritu, RF [Radio Frequency] Systems Team Leader Ramalakshmi, Solar Sub-systems Analyst Uma, Deputy Project Director for sensor systems, Padmashri S., Minal, Moumita and many others. The trends are changing now—Ramadevi T.S. retired as deputy director of the Vikram Sarabhai Space Centre, there's Mangala Mani blazing a trail in the Antarctic and women are working for Chandrayaan-2,' Roopa lists the women of ISRO flying high as she escorts me out of the building, past the gigantic antenna and the strict security. She signs off my visitor's pass. It is valid for only three hours and I have made full use of my allotted time. Next up is a visit to the ISRO headquarters, two hours away, to meet the man who selected the men *and* the women to play key roles in the Mars mission—Project Director S. Arunan.

~

At Antariksh Bhavan, the ISRO headquarters, Subbaiah Arunan reaches for the telephone to procure the list of the key team members of MOM. Out of a total seventy-four, the twenty women named make up 27 per cent of the key positions. 'But the overall composition of women in the Mars mission was higher, at around 40 per cent,' Arunan mentions. Today, after the success of the Mars Orbiter Mission, he is elated with the entire team's performance and singles out the women—in response to my queries—for liberal praise.

The thirty-minute interview spills over, as Arunan reviews the contributions of women scientists in the prestigious Mars mission. 'When my boss asked me to lead a team, I included more women in it without thinking twice, and I never regretted it. They really did their part very well, adhered to the accuracy and timeline demanded by the project, doing new things as easily as if they had been done in previous missions.

'I never had to tell anyone about their deadlines. They all knew what they had to come up with in eighteen months. It's not that I had given them five years to do it,' laughs Arunan. 'In such a project, every gram matters, every second matters, every failure is counted, always racing against time.'

I mention the unpretentiousness exhibited by the women scientists of ISRO, both in private interviews with me, and public interactions on a larger stage. Arunan calls it the ISRO culture. 'Conservatism is built into our culture. That is why we could achieve MOM with a minimum budget and short time. We explore all options to see which one ensures the least expensive, least mass- and power-consuming systems, so that overall efficiency is better,' he explains.

'What's the next big mission? Proving that men are from Venus?'

Arunan chuckles. 'The young team that we have now will take us not only to Venus but the whole solar system. These are all big challenges for the same MOM team, the same women too—Nandini and Ritu in communication, the payload team in SAC and the tracking team in ISTRAC. They have taken up higher roles, energized by the successful Mars mission. They may even lead in some missions, which will inspire women across the country. I hope to see a lady leading a challenging mission during my tenure in office here.'

Arunan shares his own learning experience from the mission, despite being a veteran. 'During the Mars mission, even though I attribute a major part of the success to the team, I myself learned a lot from Kiran Kumar, who was a review committee member for me.'

'That's exactly what the Mars mission ladies say about you. They learnt so much from their project director,' I tell him.

'See, I am still a learner. They [Kiran Kumar and K. Radhakrishnan] are mature. My experience is no match for theirs. Though all of us did a mission like this for the first time, they ensured I was not carried away, mellowing me down when needed. After all, they are answerable to the country.'

I ask him how he sees the future of ISRO, the organization in which he has spent a lifetime.

Pausing for a bit, he says, 'Today we have proved that we are second to none, having accomplished this kind of a mission in a unique way. But I would still say we are at a very nascent stage in terms of space. If you want to now focus on R&D and do much more, which is ISRO's main job, then regular things need to be outsourced to private players. In NASA, for example, even the highest advanced elements are outsourced to private parties. If ISRO goes along with Indian infrastructure, we can definitely match other countries.

'You could capitalise on the immense goodwill ISRO enjoys across India...'

'Definitely. That is my ambition—technology should come and Indian participation other than ISRO should help compliment ISRO. That is a faster way of doing things.'

~

Addressing the endless speculation about whether life on Mars will ever be possible, Ritu Karidhal is hopeful. 'Now that we are operating the payloads and getting a lot of data which is given to developers and all space physicists, a lot of papers have been published with new information. If you ask me, it should be quite possible to discover some sort of life on Mars. The planet had a very habitable environment earlier. Some two billion years ago, when the earth was just a ball of molten iron, Mars had a sufficiently thick atmosphere and water. But then it lost all this... Something must have happened. Wherever there is water and air, life is possible.'

Exploring the cosmos through more advanced missions such as Chandrayaan-2, sending an orbiter to Venus and putting a lander on Mars—all are on ISRO's space-map in the near future.

Chandrayaan-2 has an orbiter, a lander and a rover with mounted instruments to study the lunar surface and send data. Mangalyaan-2 will carry more advanced payloads for collecting data. It is expected to be more of a science mission than MOM, which was termed as a 'technology demonstrator', showing the world the technological capability of Indian space science for the design, planning, management, navigation, deep space communication and operations of an interplanetary mission.

~

Some critics panned MOM for not delivering 'exciting science' despite it being 'an engineering marvel'. SAC Director Tapan Misra says lightheartedly, 'Even though the whole country is going gaga over the Mars mission, I personally feel all of ISRO's missions are as good. Certain missions are difficult to convey to the public but Mars was familiar to people already, so we needed very little ability to express it. Woh film agar Shahrukh Khan ki chalti hai, toh storyline ki kya zaroorat hai [If a Shahrukh Khan film is successful, where's the need for a story]?'

Another ISRO director from two decades ago provides a unique, insider's take on the Mars mission and its place in ISRO's hall of fame. Eighty-year-old R. 'Dan' Aravamudan, one of the pioneering space scientists at ISRO even before its formal inception in 1969, is invited to each satellite launch. He still participates in major reviews, and was there for the pre-launch review of MOM.

Dan went to Sriharikota in November 2013 to witness the MOM launch along with his journalist-wife Gita Aravamudan. 'The Mars mission certainly attracted a lot of attention, but in terms of ISRO's total output, its developmental work and its technical content, it forms a tiny part. ISRO's main contribution is developing the capability to build satellite launch vehicles and an infrastructure for launching them. This has been done for over five decades now and *this* is a great achievement. The Mars mission is somebody's brilliant idea of using the existing vehicle, the PSLV, to optimize it for doing a very good science job. It indicates the degree of excellence of the entire endeavour. But I would say that ISRO's ability to indigenously develop a cryogenic engine, incorporated in the latest GSLV Mark 111 with a much better performance, being able to build a reliable rocket as good as any other in the world and PSLV's thirty-five successful launches are definitely a world record,' he says.

Dan counts ISRO's second significant accomplishment to be the continuous, seamless stream of indigenous scientists—many from regional, rural institutions and ordinary engineering colleges—not necessarily IITs. 'They join ISRO. They are motivated, they undertake tasks and complete them. That is the excellence of ISRO,' emphasizes the octogenarian rocket scientist.

Having worked with the legendary Vikram Sarabhai—directly reporting to him at Thumba Equatorial Rocket Launching Station (TERLS), Trivandrum, in the early sixties—does he feel that Sarabhai's vision for Indian space science and its capacity for triggering social change has been realized?

'It has been exceeded,' says Dan. 'Sarabhai had talked of building launch vehicles. Today we have the capability to launch polar and geosynchronous satellites with increased payload capacity.

'We have also done missions to the moon and Mars which weren't part of his vision. He was more focused on national and societal applications. When he talked about launching a satellite, we were in the bullock cart age. Despite knowing how dedicated he was, as were all of us young engineers then, we took it with a pinch of salt, thinking satellite launches are for big nations. But today, I see ISRO has done more than what Sarabhai dreamt.'

What about the women scientists in this larger scheme of things at ISRO, and how different is the status quo today from what it was in the early seventies?

Dan recalls his initial years at Thumba, when the entire staff comprised ten or fifteen people. Once, the principal of an engineering college invited him as an external examiner for some final year electronics engineering students. A young girl whom he quizzed about testing operational amplifiers got so tense that she fainted and fell down. She got recruited nevertheless, since she was a bright engineer. She went on to become one of the senior scientists at Thumba.

'At that time, out of say a hundred, there would be five or six women in the engineering category and they had limitations. We would call for many tours, going from one centre to another. They could not travel around and stay overnight. This would have attracted attention in the conservative society of Trivandrum. Today, there are women in significant positions as project directors, but no centre director or chairman as yet. At the recruitment level also, though women's enrollment may have improved, it probably still is not fifty-fifty. What was true decades ago is probably true even now. Between a man and a woman who are equally efficient, people choose a man. It may be an unconscious choice borne from the belief that a man will be available right through, whereas a woman will need to take care of the husband and children at home.

'In a place where you have to launch satellites and during mission times you have to be on the job twenty-four hours, travel, one has to take on people and make them work by example. There are women who do not want to be transferred from one centre to another; they prefer to stay in the background. As far as general intellectual capability is concerned, women were equally hardworking whenever they were at work and they performed well. People who performed well got promotions. The numbers are going up and the situation is improving for women scientists at ISRO,' he ends on an optimistic note.

Gita Aravamudan, a journalist and published author, adds, 'I remember some fifty-odd women from Dan's office, most of them from the administration staff, coming up to me and saying they didn't have a restroom where they could go and relax during their periods or pregnancy. I told Dan, and over a period of time they sorted it out. It's just that nobody thought of it. In my experience, ISRO was probably one of the most gender-sensitive organizations before that expression itself gained currency.'

~

A couple of years ago, while inaugurating a women's conference with 300 delegates in Thiruvananthapuram, VSSC Director and now-ISRO Chairman K. Sivan extolled the virtues of 'patience, perseverance, multi-tasking abilities, communication and analytical skills of ISRO's women employees, which brought them a fair share in the organization's success.' However, as long as the triumphs of women scientists are acknowledged and lauded mainly in women's day celebrations, women's conferences and women's awards, there isn't any true gender parity. If women are limited to presenting papers at a Women's Science Congress instead of plenary sessions at the larger Indian Science Congress, then women panelists are well within their rights to ask for this tokenism to be scrapped. As well-respected physicist Rohini Godbole once described pithily, 'Diversity is the root of all science. If it doesn't include half of humanity, then it is science that also suffers.'

And yet, women achievers in science need to be publicly recognized and feted a lot more than they are, because they are game-changers in reducing the gap decade by decade, generation by generation. Change *is* on its way even if it is a little slow, a little onerous, a little too little. Any lasting change will occur when the mindset of an entire generation changes. When scoffing boys—who grow up to become chauvinist men—stop believing that 'science is not for girls' or that women treat their jobs as a hobby or stop-gap before marriage and family. The next generation of intrepid young women searching for their own glory have some of the road laid out for them by ISRO's women scientists.

These scientists come from modest, conservative backgrounds, from small places like Sagar, Dehradun or Ariyalur, having studied in vernacular medium schools.

Those Magnificent Women and Their Flying Machines

Many of them overcame parental and societal opposition to follow their inner voice. They ended up leading missions such as RISAT-1 or GSAT-12, or landing key roles in MOM. As Vellarmathi claims, they have proved to be 'useful to the nation'. Handling high-pressure assignments, which call for pinpoint precision according to tightly planned schedules, these women have come up with out-of-the-box concepts at work while also supervising family duties. They have done this with a minimum of fuss and fanfare.

The only hint at the gender imbalance faced by them lies in their unconscious use of expressions such as 'adjustment', 'working harder to balance roles' or painstakingly crediting their families and husbands for their 'understanding and support'. Sometimes, they have even given up on the American dream or a different career to follow in their husband's footsteps—to the job or city of his choice.

For every man who thinks that women are meant to primarily look after their families and tend to children, there is a Ritu Karidhal, who works three shifts to launch interplanetary satellites, while her husband looks after the home front. Or a Durga Darshini who takes just two days off to get married and flies back to complete a mission. For every boss who feels maternity leave will not allow a woman to take on rigorous tasks, there is a Durga Darshini who goes directly from her office to the hospital for her child's delivery. For every chauvinist colleague who says that women cannot work 24x7 or travel at short notice, there is a Seetha Somasundaram who stayed up nights as the only woman to conduct optical astronomy experiments. Or a Nandini Harinath, Minal Sampath or Moumita Dutta, along with several others, who do not go home for days in a row during launch times, and clock in miles of air travel during preparations. For every old boys' club that passes along inappropriate jokes or attempts to mansplain, there is an Arundhati Misra who shuts it down effectively.

Breaking boundaries in the male-dominated sphere of space science may not be easy, but these serving scientists go about it in an orderly and understated fashion, doing whatever works for them. Nandini Harinath advises girls to not get unduly bogged down by disappointments. Anuradha T.K. advocates being proud of what you do and sharing your work with your family. Shahana Khaleel recommends finding different ways to get work done, by not limiting oneself.

ISRO's emphasis on team effort also helps, and triumphs are never considered the result of personal contributions, male or female. On this subject, Ritu Karidhal cites Mitch Albom's story about the first wave of an ocean getting frightened of crashing against the shore and getting washed away. 'The second wave tells the first, "You are scared because you see yourself only as a wave. I am not scared because I see myself as a part of the ocean that cannot be erased." ISRO-ites are cohesive like the second wave, working together towards their goals.'

~

The Satish Dhawan Space Centre in Sriharikota is my last stop in the story about ISRO's superwomen. It is the place where dreams are launched in the form of rockets. This is where Anuradha T.K. and her all-women team launched GSAT-12 into its geosynchronous orbit. This is where RISAT-1, launched by the team led by N. Vellarmathi, started its ascent into the skies. This was also MOM's first take-off point.

The SDSC, named after former ISRO chairman Satish Dhawan, was originally known as the Sriharikota Range. It was here that one of the first ISRO rockets, Rohini-125, was tested as early as 9 October 1971.

From that day onwards, there has been no looking back. A total of ninety-seven spacecraft missions, sixty-seven launch missions and 237 foreign satellites have been launched till date, with a lot more waiting to take off.

Sriharikota is not an easy place to visit. In fact, unless you have prior permission, you cannot get any access to it. Accredited media persons, a select number of guests chosen to attend a satellite launch and ISRO staff members are the only people allowed to enter the heavily-guarded premises of SDSC.

~

On a sweltering August day, I join the throng of visitors and media personnel in the large, air-conditioned buses that are headed for the launch centre at Sriharikota. Three hours later, the bus veers off the main highway, vending its way through a narrower road flanked by wetlands on both sides. We cross the vast expanse of uninhabited area—a pre-requisite for ensuring a safety zone near a spaceport—and drive through two armed checkpoints to the Media Centre inside SDSC. Two tall, scaled models of the PSLV and GSLV rockets grace the entrance. After presenting my visitor's pass for the third time, I make my way inside the garish, mustard-yellow building, past the video cameras and press photographers in the lobby, and enter the vast auditorium on the first floor.

Rows upon rows of desks and chairs face the three gigantic screens that form the backdrop to the array of mikes for the post-launch press conference. The screens show the ISRO scientists at the SDSC mission control room, with the words, 'PSLV-C39 IRNSS-IH mission, Countdown in progress' labelled below. I can spot a few women among the staff huddled over their computer screens in the video. There are just a few minutes left for the navigation satellite to take off from the second launch pad, located approximately 8 km from the Media Centre. Then the entire auditorium empties out and everyone flocks to the terrace two floors above to witness the launch first-hand.

In one corner of the sprawling terrace, ISRO Public Relations Officer, B.R. Guruprasad, explains the four stages of rocket separation to the waiting media. Meanwhile, people like me scan the darkening skies above, searching for the best possible lookout. A small moon is already visible in the crimson-streaked hues of the evening.

The sound of crickets is soon overtaken by the stentorian commands by multiple male voices over the public address system. The technical queries and their incomprehensible replies set up an air of hushed expectancy. A fresh voice announces, 'T minus 1 minute 27 seconds,' but the chattering on the terrace continues, albeit at a lower decibel level. At one minute to lift-off, a female voice takes over the counting: T minus 30 seconds, T minus 25 seconds and so on. At 6.59 pm, the sky is completely dark, making for an easier viewing.

The instructions continue and as the seconds are counted off, there is complete silence on the terrace. When ten seconds are left, the numbers are ticked off in quick succession. At T0, an aura of incandescent light rises up behind the canopy of trees not too far from where I am standing. The countdown continues—plus 2, plus 4, plus 7—even as the light spreads wide and high, illuminating the entire sky in a spectacular arc. 'Lift-off normal,' says the earlier male voice. He counts till plus 25, at which point another announcement is made: 'First stage performance: normal.'

In this short time, a little over sixty seconds, the luminescent beam narrows into a great flaming ball of fire that keeps climbing higher. In less than a heartbeat the patterns of overhead clouds knotted together are back-lit, even as the round orb of light becomes smaller and smaller, gradually vanishing into the darkness. The whole sequence feels like watching the sun rise from the earth at a fast-forward pace and vanish into another cosmos.

As total darkness envelops the terrace once again, I can hear the renewed sound of clapping and cheers all around me. In the space of a minute, the spectacle transported me to another world. It takes me another minute to realize that this entire phenomenon was man-made.

Crossing the Rubicon

I imagine a young girl, somewhere in the country, watching the same spectacle on a television screen—perhaps sitting in a village community hall—visualizing a day when *she* can blaze a trail in space as high as the rocket soaring in front of her, into the great unknown.

Epilogue

Following is a list of developments related to ISRO's work, which have taken place after the events of this book:

1. A huge lake of liquid water was found just below the surface of Mars, by astronomers using the Mars Advanced Radar for Subsurface and Ionosphere Sounding (MARSIS) onboard the Mars Express Orbiter.[60]
2. The one-year trial phase of ISRO's satellite-based warning system at unmanned railway crossings proved satisfactory in May 2018. The hooter warns users at unmanned crossings, becoming louder as the train nears a distance of 500 m–4 km, and is silent after passing. Unmanned crossings are a major hazard, with 5,792 unmanned crossings till date accounting for 40 per cent accidents involving the railways.
3. In a new initiative undertaken with the Armed Medical Services in August 2018, ISRO's telemedicine programme connecting remote/rural medical colleges and hospitals to specialty hospitals in cities through Indian satellites, will now reach combat soldiers, airmen and sailors deployed in isolated posts such as the Siachen glacier. ISRO's telemedicine network covers 384 hospitals, with sixty specialty hospitals beamed via INSAT-3A and GSAT-12 satellites to 306

Epilogue

remote rural/district medical college hospitals and eighteen mobile telemedicine units across India.

4. ISRO TV, a space science channel telecasting programmes about the organization's missions and applications, especially aimed at developing scientific rigour among children and youth, will soon be operational.
5. The rocket launch pad at Sriharikota will be opened to the public, allowing the common man to watch missions take off into space. Space museums will also be opened across the country to educate and inform people about space technologies and advances in the sector.
6. On 15 August 2018, Independence Day, Prime Minister Narendra Modi promised that a 'son or a daughter' would take the Indian flag to outer space by 2022. A woman is leading ISRO's first human space flight mission, Gaganyaan—the 56-year-old Control Systems Engineer Dr Lalithambika. Gaganyaan will attempt to send three Indians into the low earth orbit for up to seven days.[61]
7. Two unmanned flights using a humanoid (a robot resembling a human) are planned for December 2020 and July 2021.[62]
8. Astrosat, India's first dedicated multi wavelength space observatory, completed three years in September 2018, having observed over 750 sources so far and provided data for almost 100 publications in peer-reviewed journals.
9. The Mars Orbiter or Mangalyaan continues to

circle Mars, sending pictures and data far beyond its expected mission life of six months. At the time of its fourth anniversary in September 2018, the Mars Colour Camera had captured more than 980 images.

10. NASA's Insight landed on Mars on 27 November 2018, after seven months of travel covering a distance of 483 million km. This unmanned lander will remain on the surface of Mars for two years, providing information about marsquakes as well as the earth's origins. Insight captured raw sounds on Mars and sent eleven photographs of the Lander and the ground around it.

11. The total tally of ISRO's missions till date is 101 spacecraft missions, including three nano satellites and one micro satellite; seventy-one launch missions, nine student satellites, two re-entry missions and 269 foreign satellites (of thirty-two countries).

Acknowledgements

The time has come to talk of many things, of the Himalayas and home—and the confining desk—of enablers and friends.

Six snow-capped Himalayan peaks at Sonapani kicked off the first chapter of this book amidst the quiet hospitality of Himalayan village friends, Deepa and Ashish Arora.

The second, much longer stint at my Pune home ensured a pampered existence, free of any kind of work or worry, as always.

For facilitating the research, travel and keeping the body and soul intact for eighteen months—from the stirrings of an idea to the final manuscript—Rajkamal Vempati deserves the biggest bow of them all.

My heartfelt thanks to all these magnificent people of *my* universe.

Notes

Prologue

1. "Let Your Child Reach the Stars," YouTube, https://www.youtube.com/watch?v=ODWrTVSFcGs.
2. "Quick Take: Women in Science, Technology, Engineering, and Mathematics (STEM)," Catalyst, accessed on February 3, 2018, https://www.catalyst.org/knowledge/women-science-technology-engineering-and-mathematics-stem.
3. Ibid.
4. Desai and Jameel, "Women of Science,' *The Indian Express*, October 9, 2017, https://indianexpress.com/article/opinion/columns/scientific-research-in-india-women-scientiests-women-of-science-4881037/.
5. Reshma Ganeshbabu, "Nobel Prize Winners: What's the count of women laureates," ShethePeople.tv, October 13, 2017, https://www.shethepeople.tv/blog/nobel-prize-winners-count-women-laureates.
6. Jason McBride, "Nobel laureate Donna Strickland: 'I see myself as a scientist, not a woman in science," *The Guardian*, October 20, 2018, https://www.theguardian.com/science/2018/oct/20/nobel-laureate-donna-strickland-i-see-myself-as-a-scientist-not-a-woman-in-science.
7. Desai and Jameel, "Women of Science."
8. Rohini M. Godbole and Ramakrishna Ramaswamy, "Women Scientists in India," accessed on September 5, 2017, https://www.ias.ac.in/public/Resources/Initiatives/Women_in_Science/AASSA_India.pdf.
9. Abhishek Chari, Shruti Murlidhar and Navneet A. Vasistha, "Dear Nobel Laureate, Your Words Matter

Because Young Women Are Listening to You," *The Wire*, February 7, 2018, https://thewire.in/science/dear-nobel-laureate-words-matter-young-women-listening.
10. "Science Career for Indian Women," New Delhi: Indian National Science Academy, 2004, accessed on October 7, 2017, https://www.ias.ac.in/public/Resources/Initiatives/Women_in_Science/report.pdf.

1. Women From Mars

11. Michael Irving, "Huge lake of liquid water found on Mars," *New Atlas*, July 25, 2018, https://newatlas.com/liquid-water-lake-found-mars/55592/.
12. "Fun Facts About Mars." NASA, accessed on July 13, 2017. https://nasa.tumblr.com/post/141602045589/fun-facts-about-mars.
13. P.V. Rao Manoranjan, B.N. Suresh, and V.P. Balagangadharan, eds. *From Fishing Hamlet to Red Planet: India's Space Journey* (Delhi: Harper Collins, 2015), 565.
14. Shakoor Rather, "Over 1.3 lakh Indians 'book ticket' to Mars," *Livemint*, November 16, 2017, https://www.livemint.com/Science/q8AngnaSgfbvsbT06Hdp1K/Over-13-lakh-Indians-book-ticket-to-Mars.html.
15. "100,000 people apply to go to Mars and not return." *Times of India*, August 11, 2013. https://timesofindia.indiatimes.com/home/science/100000-people-apply-to-go-to-Mars-and-not-return-Project/articleshow/21757877.cms.
16. Don Reisinger, "Elon Musk's Tesla Missed Mars Orbit After Successful Falcon Heavy Launch," *Fortune*, February 7, 2018, http://fortune.com/2018/02/07/elon-musk-tesla-mars-orbit/.
17. Rafi Letzter, "Nasa Has a Plan to Put Robot Bees On Mars," *LiveScience*, April 3, 2018, https://www.

livescience.com/62204-nasa-plans-robot-bees-on-mars.html.
18. "Welles scares nation," *History*, accessed on August 19, 2017. https://www.history.com/this-day-in-history/welles-scares-nation.
19. *From Fishing Hamlet to Red Planet: India's Space Journey*, 621.
20. A payload is the carrying capacity of a launch vehicle, usually measured in terms of weight. The payload can take the form of instruments such as cameras, telescopes, cargo and even a human being in a manned mission.
21. "Mars Orbiter Mission—Spacecraft & Mission Overview," *Spaceflight 101*, accessed on June 27, 2017, http://www.spaceflight101.net/mars-orbiter-mission.html.
22. K. Radhakrishnan and Nilanjan Routh, *My Odyssey* (Delhi: Penguin Random House, 2016), 260.
23. "Budget at a Glance," ISRO, accessed on February 15, 2018, https://www.isro.gov.in/sites/default/files/AnnualReports/2014/images/BG1.jpg.
24. D. Balasubramanian. "Mangalyaan: a steal at Re. 4 per person," *The Hindu*, November 13, 2013, https://www.thehindu.com/sci-tech/science/mangalyaan-a-steal-at-re4-per-person/article5347627.ece.
25. Ipsita Agarwal, "These Scientists Sent a Rocket to Mars for Less Than It Cost to Make 'The Martian'," *Wired*, March 17, 2017. https://www.wired.com/2017/03/these-scientists-sent-a-rocket-to-mars-for-less-than-it-cost-to-make-the-martian/.
26. As per a document listing key MOM team members shared by S. Arunan.
27. "Wikipedia: Indian Women in Science Edit-a-thon," Wikipedia, accessed on July 19, 2017, https://en.wikipedia.org/wiki/Wikipedia:Indian_Women_in_Science_Edit-a-thon.

Notes

2. MOM: Operations in Outer Space

28. Ashwaq Masoodi, "Where are India's female scientists?". *Livemint*, April 19, 2016, https://www.livemint.com/Politics/N0j16kL2ZAzk94ssOKoTdK/Where-are-Indias-female-scientists.html.
29. The assembly, integration and testing of the satellite's payloads and sub-systems take place in a vast 'clean room', which as the name suggests, is a completely airlocked chamber. Its temperatures range from 1° C to 22° C and a cleanliness level of 1,00,000 particles permissible per cubic metre.
30. Godbole and Ramaswamy, "Women Scientists in India."
31. Masoodi, "Where are India's female scientists," 2016.
32. Paul Voosen, "Women make up just 15% of NASA's planetary mission science teams. Here's how the agency is trying to change that," *Science*, May 4, 2017. https://www.sciencemag.org/news/2017/05/women-make-just-15-nasa-s-planetary-mission-science-teams-here-s-how-agency-trying.
33. Godbole and Ramaswamy. "Women Scientists in India."
34. "Science Career for Indian Women," 12.
35. Anitha Kurup, Kantharaju B., Maithreyi R. and Rohini Godbole, "Trained Scientific Women Power," (Bengaluru: Indian Academy of Sciences-National Institute of Advanced Studies, 2010).
36. "SRE-1," Department of Space: ISRO, January 10, 2007, https://www.isro.gov.in/Spacecraft/sre-1-0.
37. "Space Capsule Successfully Recovered," Department of Space: ISRO, January 22, 2007, https://www.isro.gov.in/update/22-jan-2007/space-capsule-successfully-recovered.

3. MOM: The Payload Performers

38. In 2018 Misra was appointed as an adviser to ISRO Chairman K. Sivan.

39. The five payloads in MOM were the Methane Sensor for Mars (MSM), the Mars Color Camera (MCC), the Therma Infrared Imaging Spectrometer (TIS), the Lyman-Alpha Photometer (LAM) and the Mars Exospheric Neutral Composition Analyser (MENCA).
40. INSAT-3D is an advanced meteorological, data relay and satellite-aided search and rescue satellite, launched on July 26, 2013. Its major users in India are the Coast Guard, Airports Authority of India, defence services and fishermen.
41. In a satellite like the Mars Orbiter, sub-systems include structural systems, thermal and digital, sensors, power, communication, payload, the launch vehicle, spacecraft propulsion, remote sensing and navigation. These are further categorized into sub-system elements. E.g., transmitters, receivers and antennae are the elements of the communication sub-system of the space satellite.

4. The Vanguard Veterans

42. Gamma Ray Bursts (GRBs) are extremely energetic explosions that have been observed in distant galaxies. They are the brightest electromagnetic events known to occur in the universe and can last from ten milliseconds to several hours.
43. *From Fishing Village to Red Planet: India's Space Journey*, 128.
44. Vandana Singh, "DST's Initiatives for Women Scientists and Technologists", Science India Fest, May, 2018, https://scienceindiafest.org/wp-content/uploads/2018/05/Dr.VandanaSingh-DST-WSE.pdf.
45. The RISAT-1 completed five years of its scheduled operations in 2017 and is now no longer functional.
46. A satellite bus is the infrastructure of the spacecraft, usually the locations of the payloads.

47. The moon mission, Chandrayaan-2 has been announced by ISRO. See epilogue.
48. Subodh Varma, "Fewer working women in India than Nepal", *Times of India*, September 17, 2017, https://timesofindia.indiatimes.com/india/fewer-working-women-in-india-than-nepal/articleshow/60715980.cms
49. "2.39 lakh girls under five die in India every year because of gender discrimination, finds study." Scroll, May 15, 2018, https://scroll.in/latest/879054/2-39-lakh-girls-under-five-die-in-india-every-year-because-of-gender-discrimination-finds-study.

5. Beyond MOM: The Applications Achievers

50. The highest post at ISRO is the Chairman, followed by Directors of all centres and the Deputy Directors. Then come Distinguished Scientist I and II, Scientist/Engineer H, G, SG, SF, SE, SD and the entry level SC.
51. A weather satellite also used for search and rescue operations, launched in September 2016.
52. These are nicknames for the latitude bands that span 40–50 degrees, 50–60 degrees and below 60 degrees in the Southern Ocean—the extreme southern portion of the Pacific, Indian and Atlantic oceans. These are areas of strong westerly winds and ferocious storms.
53. Stratigraphy is a branch of geology concerned with the study of rock layers or strata and layering.
54. Sea-ice is the ice that forms as seawater freezes. Since ice is less dense than water, it floats on the ocean's surface.
55. A scatterometer is an active microwave sensor/device that measures wind speed and direction, determining ocean surface level wind vectors
56. Scatsat-1 is a state-of-the-art satellite with a dedicated Scatterometer payload to provide wind vector data products for weather forecasting, cyclone detection and tracking services for users.

57. Earlier, before Cyclone Phailin struck Odisha, Andhra Pradesh and Bihar in October 2013, more than 5,50,000 people were evacuated to safety based on the warning data given by the ISRO satellites.
58. The National Remote Sensing Centre is an ISRO centre responsible for remote sensing data acquisition and processing, dissemination of data, aerial remote sensing and decision support for disaster management.

6. Crossing The Rubicon

59. Due to the Mars-earth-sun geometry, the Orbiter was in eclipse behind Mars and dependent solely on battery power. The radio link between the Orbiter and the earth was blocked by Mars, and there was a communication blackout with telemetry resuming after the burn ended.

Epilogue

60. Eric Betz, "Vast lake of liquid water discovered on Mars," *Astronomy*, July 25, 2018, http://www.astronomy.com/news/2018/07/liquid-water-on-mars.
61. Surendra Singh, "Rs 10,000 crore plan to send 3 Indians to space by 2022," *Times of India*, December 29, 2018, https://timesofindia.indiatimes.com/india/union-cabinet-clears-rs-10000cr-for-indias-gaganyaan-project/articleshow/67288124.cms.
62. Chethan Kumar, "Before humans, 'humanoids' to do experiments in space," *Times of India*, January 18, 2019, https://timesofindia.indiatimes.com/india/before-humans-humanoids-to-do-experiments-in-space/articleshow/67580180.cms.

References

"2.39 lakh girls under five die in India every year because of gender discrimination, finds study." Scroll. May 15, 2018. https://scroll.in/latest/879054/2-39-lakh-girls-under-five-die-in-india-every-year-because-of-gender-discrimination-finds-study.

"Budget at a Glance." ISRO. Accessed on February 15, 2018. https://www.isro.gov.in/sites/default/files/AnnualReports/2014/images/BG1.jpg.

"Fun Facts About Mars." NASA. Accessed on July 13, 2017. https://nasa.tumblr.com/post/141602045589/fun-facts-about-mars.

"Mars Orbiter Mission—Spacecraft & Mission Overview." Spaceflight 101. Accessed on June 27, 2017. http://www.spaceflight101.net/mars-orbiter-mission.html.

"Quick Take: Women in Science, Technology, Engineering, and Mathematics (STEM)," Catalyst. Accessed on February 3, 2018. https://www.catalyst.org/knowledge/women-science-technology-engineering-and-mathematics-stem.

"Science Career for Indian Women." New Delhi: Indian National Science Academy, 2004. Accessed on October 7, 2017. https://www.ias.ac.in/public/Resources/Initiatives/Women_in_Science/report.pdf.

"Let Your Child Reach the Stars." YouTube. Accessed October 17, 2017. https://www.youtube.com/watch?v=ODWrTVSFcGs.

"Space Capsule Successfully Recovered." Department of Space: ISRO. January 22, 2007. https://www.isro.gov.in/update/22-jan-2007/space-capsule-successfully-recovered.

"SRE-1." Department of Space: Indian Space Research Organisation. January 10, 2007. https://www.isro.gov.in/Spacecraft/sre-1-0.

"Welles scares nation." *History*. Accessed on August 19, 2017. https://www.history.com/this-day-in-history/welles-scares-nation.

"Wikipedia: Indian Women in Science Edit-a-thon." Wikipedia. Accessed on July 19, 2017. https://en.wikipedia.org/wiki/Wikipedia:Indian_Women_in_Science_Edit-a-thon.

"100,000 people apply to go to Mars and not return." *Times of India*. August 11, 2013. https://timesofindia.indiatimes.com/home/science/100000-people-apply-to-go-to-Mars-and-not-return-Project/articleshow/21757877.cms.

Agarwal, Ipsita. "These Scientists Sent a Rocket to Mars for Less Than It Cost to Make 'The Martian'." *Wired*. March 17, 2017. https://www.wired.com/2017/03/these-scientists-sent-a-rocket-to-mars-for-less-than-it-cost-to-make-the-martian/.

Balasubramanian, D. "Mangalyaan: a steal at Re.4 per person." *The Hindu*. November 13, 2013. https://www.thehindu.com/sci-tech/science/mangalyaan-a-steal-at-re4-per-person/article5347627.ece.

Betz, Eric. "Vast lake of liquid water discovered on Mars." *Astronomy*, July 25, 2018, http://www.astronomy.com/news/2018/07/liquid-water-on-mars.

Chari, Abhishek, Shruti Murlidhar and Navneet A. Vasistha. "Dear Nobel Laureate, Your Words Matter Because Young Women Are Listening to You." The Wire. February 7, 2018. https://thewire.in/science/dear-nobel-laureate-words-matter-young-women-listening.

Desai and Jameel. "Women of Science.' *The Indian Express*, October 9, 2017, https://indianexpress.com/article/opinion/columns/scientific-research-in-india-women-scientiests-women-of-science-4881037/.

Godbole, Rohini M., Ramakrishna Ramaswamy. "Women Scientists in India." Accessed on September 5, 2017. https://

References

www.ias.ac.in/public/Resources/Initiatives/Women_in_Science/AASSA_India.pdf.

Irving, Michael. "Huge lake of liquid water found on Mars." New Atlas. July 25, 2018. https://newatlas.com/liquid-water-lake-found-mars/55592/.

Kumar, Chethan. "Before humans, 'humanoids' to do experiments in space," *Times of India*, January 18, 2019, https://timesofindia.indiatimes.com/india/before-humans-humanoids-to-do-experiments-in-space/articleshow/67580180.cms.

Kurup, Anitha, Kantharaju B., Maithreyi R., Rohini Godbole. "Trained Scientific Women Power." Bengaluru: Indian Academy of Sciences-National Institute of Advanced Studies, 2010.

Letzter, Rafi. "Nasa Has a Plan to Put Robot Bees On Mars." *LiveScience*. April 3, 2018. https://www.livescience.com/62204-nasa-plans-robot-bees-on-mars.html.

Masoodi, Ashwaq. "Where are India's female scientists." *Livemint*. April 19, 2016. https://www.livemint.com/Politics/N0j16kL2ZAzk94ssOKoTdK/Where-are-Indias-female-scientists.html.

McBride, Jason. "Nobel laureate Donna Strickland: 'I see myself as a scientist, not a woman in science," *The Guardian*. October 20, 2018, https://www.theguardian.com/science/2018/oct/20/nobel-laureate-donna-strickland-i-see-myself-as-a-scientist-not-a-woman-in-science.

Radhakrishnan, K., Nilanjan Routh. *My Odyssey*. Delhi: Penguin Random House, 2016.

Rao, P.V. Manoranjan, B.N. Suresh, and V.P. Balagangadharan, eds. *From Fishing Hamlet to Red Planet: India's Space Journey*. Delhi: HarperCollins, 2015.

Rather, Shakoor. "Over 1.3 lakh Indians 'book ticket' to Mars." *Livemint*. November 16, 2017. https://www.livemint.com/Science/q8AngnaSgfbvsbT06Hdp1K/Over-13-lakh-Indians-book-ticket-to-Mars.html.

Reisinger, Don. "Elon Musk's Tesla Missed Mars Orbit After Successful Falcon Heavy Launch." *Fortune*. February 7, 2018. http://fortune.com/2018/02/07/elon-musk-tesla-mars-orbit/.

Reshma Ganeshbabu. "Nobel Prize Winners: What's the count of women laureates." ShethePeople.tv. October 13, 2017. https://www.shethepeople.tv/blog/nobel-prize-winners-count-women-laureates.

Singh, Surendra. "Rs 10,000 crore plan to send 3 Indians to space by 2022." *Times of India*. December 29, 2018. https://timesofindia.indiatimes.com/india/union-cabinet-clears-rs-10000cr-for-indias-gaganyaan-project/articleshow/67288124.cms.

Varma, Subodh. "Fewer working women in India than Nepal." *Times of India*. September 17, 2017. https://timesofindia.indiatimes.com/india/fewer-working-women-in-india-than-nepal/articleshow/60715980.cms.

Voosen, Paul. "Women make up just 15% of NASA's planetary mission science teams. Here's how the agency is trying to change that." *Science*. May 4, 2017. https://www.sciencemag.org/news/2017/05/women-make-just-15-nasa-s-planetary-mission-science-teams-here-s-how-agency-trying.

www.ingramcontent.com/pod-product-compliance
Lightning Source LLC
Chambersburg PA
CBHW052049220426
43663CB00012B/2504